*The Palgrave Macmillan Animal Ethics Series*

Series editors: **Andrew Linzey** and **Priscilla Cohn**

In recent years, there has been a growing interest in the ethics of our treatment of animals. Philosophers have led the way and now a range of other scholars has followed, from historians to social scientists. From being a marginal issue, animals have become an emerging issue in both ethics and multidisciplinary inquiry. This series explores the challenges that Animal Ethics poses, both conceptually and practically, to traditional understandings of human–animal relations.

Specifically, the Series will:

- provide a range of key introductory and advanced texts that map out ethical positions on animals;
- publish pioneering work written by new, as well as accomplished, scholars; and
- produce texts from a variety of disciplines that are multidisciplinary in character or have multidisciplinary relevance.

*Titles include*:

AN INTRODUCTION TO ANIMALS AND POLITICAL THEORY
Alasdair Cochrane

THE COSTS AND BENEFITS OF ANIMAL EXPERIMENTS
Andrew Knight

POPULAR MEDIA AND ANIMAL ETHICS
Claire Molloy

ANIMALS, EQUALITY AND DEMOCRACY
Siobhan O'Sullivan

ANIMALS AND SOCIAL WORK: A MORAL INTRODUCTION
Thomas Ryan

AN INTRODUCTION TO ANIMALS AND THE LAW
Joan Schaffner

*Forthcoming titles*:

HUMANS AND ANIMALS: THE NEW PUBLIC HEALTH PARADIGM
Aysha Akhtar

HUMAN ANIMAL RELATIONS: THE OBLIGATION TO CARE
Mark Bernstein

ANIMAL ABUSE AND HUMAN AGGRESSION
Eleonora Gullone

ANIMALS IN THE CLASSICAL WORLD: ETHICAL PERCEPTIONS
Alastair Harden

POWER, KNOWLEDGE, ANIMALS
Lisa Johnson

AN INTRODUCTION TO ANIMALS AND SOCIOLOGY
Kay Peggs

**The Palgrave Macmillan Animal Ethics Series**
**Series Standing Order ISBN 978–0–230–57686–5 Hardback**
**978–0–230–57687–2 Paperback**
(*outside North America only*)

You can receive future titles in this series as they are published by placing a standing order. Please contact your bookseller or, in case of difficulty, write to us at the address below with your name and address, the title of the series and the ISBN quoted above.

Customer Services Department, Macmillan Distribution Ltd, Houndmills, Basingstoke, Hampshire RG21 6XS, UK

# Animals and Social Work: A Moral Introduction

Thomas Ryan

First published 2011 by
PALGRAVE MACMILLAN

Palgrave Macmillan in the UK is an imprint of Macmillan Publishers Limited, registered in England, company number 785998, of Houndmills, Basingstoke, Hampshire RG21 6XS.

Palgrave Macmillan in the US is a division of St Martin's Press LLC, 175 Fifth Avenue, New York, NY 10010.

Palgrave Macmillan is the global academic imprint of the above companies and has companies and representatives throughout the world.

Palgrave® and Macmillan® are registered trademarks in the United States, the United Kingdom, Europe and other countries.

ISBN 978-0-230-27250-7 hardback

This book is printed on paper suitable for recycling and made from fully managed and sustained forest sources. Logging, pulping and manufacturing processes are expected to conform to the environmental regulations of the country of origin.

A catalogue record for this book is available from the British Library.

Library of Congress Cataloging-in-Publication Data
Ryan, Thomas, 1958 June 25–
    Animals and social work : a moral introduction / Thomas Ryan.
        p.   cm.
    Includes index.
    ISBN 978-0-230-27250-7 (hardback)
    1. Animal welfare—Moral and ethical aspects.   2. Animal rights—Moral and ethical aspects.   3. Social service.   I. Title.
    HV4708.R925 2011
    179'.3—dc22                                                    2011008048

10   9   8   7   6   5   4   3   2   1
20   19   18   17   16   15   14   13   12   11

Printed and bound in Great Britain by
CPI Antony Rowe, Chippenham and Eastbourne

To my grandmother Mary Murtagh, who nurtured a love of the natural world and the written word; my parents Tom and Catherine; my wife Blanca and our children Thomas-Liam, Jude, Immogen and Mirabehn; and our four-legged family members, Tess, Lucy, Simone, Jayke and Clarabelle. They have all blessed me with their unconditional love, and it is in their shelter that I live.

As I love nature, as I love singing birds, and gleaming stubble, and flowing rivers, and morning and evening, and summer and winter, I love thee.

Henry David Thoreau (1980, p. 285)

And the trees & birds & beasts & men behold their eternal joy!
... for every thing that lives is holy!

William Blake (1941, p. 200)

> Therefore am I still
> A lover of the meadows and the woods,
> And mountains; and of all that we behold
> From this green earth; of all the mighty world
> Of eye and ear, both what they half create,
> And what perceive; well pleased to recognise
> In nature and the language of the sense,
> The anchor of my purest thoughts, the nurse,
> The guide, the guardian of my heart, and soul
> And all my moral being.

William Wordsworth (n.d., p. 116)

# Contents

# Series Preface

This is a new book series for a new field of inquiry: Animal Ethics.

In recent years there has been a growing interest in the ethics of our treatment of animals. Philosophers have led the way, and now a range of other scholars have followed from historians to social scientists. From being a marginal issue, animals have become an emerging issue in ethics and in multidisciplinary inquiry.

In addition, a rethink of the status of animals has been fuelled by a range of scientific investigations that have revealed the complexity of animal sentiency, cognition and awareness. The ethical implications of this new knowledge have yet to be properly evaluated, but it is becoming clear that the old view that animals are mere things, tools, machines or commodities cannot be sustained ethically.

But it is not only philosophy and science that are putting animals on the agenda. Increasingly, in Europe and the United States, animals are becoming a political issue as political parties vie for the 'green' and 'animal' vote. In turn, political scientists are beginning to look again at the history of political thought in relation to animals, and historians are beginning to revisit the political history of animal protection.

As animals grow as an issue of importance, so there have been more collaborative academic ventures leading to conference volumes, special journal issues, indeed new academic animal journals as well. Moreover, we have witnessed the growth of academic courses, as well as university posts, in Animal Ethics, Animal Welfare, Animal Rights, Animal Law, Animals and Philosophy, Human–Animal Studies, Critical Animal Studies, Animals and Society, Animals in Literature, Animals and Religion – tangible signs that a new academic discipline is emerging.

'Animal Ethics' is the new term for the academic exploration of the moral status of the non-human – an exploration that explicitly involves a focus on what we owe animals morally, and which also helps us to understand the influences – social, legal, cultural, religious and political – that legitimate animal abuse. This series explores the challenges that Animal Ethics poses, both conceptually and practically, to traditional understandings of human–animal relations.

The series is needed for three reasons: (1) to provide the texts that will service the new university courses on animals; (2) to support the increasing number of students studying and academics researching in animal-related fields; and (3) because there is currently no book series that is a focus for multidisciplinary research in the field.

Specifically, the series will

- provide a range of key introductory and advanced texts that map out ethical positions on animals;
- publish pioneering work written by new, as well as accomplished, scholars; and
- produce texts from a variety of disciplines that are multidisciplinary in character or have multidisciplinary relevance.

The new Palgrave Macmillan Series on Animal Ethics is the result of a unique partnership between Palgrave Macmillan and the Ferrater Mora Oxford Centre for Animal Ethics. The series is an integral part of the mission of the Centre to put animals on the intellectual agenda by facilitating academic research and publication. The series is also a natural complement to one of the Centre's other major projects, the *Journal of Animal Ethics*. The Centre is an independent 'think tank' for the advancement of progressive thought about animals, and is the first Centre of its kind in the world. It aims to demonstrate rigorous intellectual inquiry and the highest standards of scholarship. It strives to be a world-class centre of academic excellence in its field.

We invite academics to visit the Centre's website, www.oxfordanimal ethics.com and to contact us with new book proposals for the series.

**Andrew Linzey and Priscilla N. Cohn**
**General Editors**

# Acknowledgements

The author and the publisher gratefully acknowledge permission to include the following copyright material:

Gerard Manley Hopkins, lines from 'Pied Beauty', in *Hopkins: Selections*, edited by Graham Storey (Oxford University Press, 1967), by permission of Oxford University Press.

Mary Midgley, from *Beast and Man: The Roots of Human Nature* (Routledge, 1996), by permission of Taylor & Francis Books (UK).

Iris Murdoch, from *The Sovereignty of Good* (Routledge, 1996), by permission of Taylor & Francis Books (UK).

Judith Wright, line from 'Birds', in *Birds: Poems* (National Library of Australia, 2004), by permission of ETT Imprint.

I express my heartfelt gratitude to Pauline Meemeduma for her perspicacity, generosity and belief, and to Andrew Linzey for his persistence and encouragement, without which this book would not have come to fruition. I shall be ever grateful to my sons Thomas-Liam and Mirabehn for saving me from countless computer related disasters. Finally, I'm indebted to Priya Venkat for her prompt and courteous assistance throughout the copy editing process.

# 1
# Animals, Social Work and the Western Tradition

> For that which befalleth the sons of man befalleth the beasts.
> Even one thing befalleth them: as the one dieth, so dieth the
> other; yea they all have one breath, so that a man has no pre-
> eminence above a beast, for all is vanity.
>
> *Ecclesiastes* 3: 19.

A social worker receives a phone call from a woman who discloses that she is a victim of domestic violence. Although she and her husband separated some six months earlier, he has continued to routinely harass and make all manner of threats against her. On attending the woman's home, the social worker learns that the violence is longstanding and severe in nature, and that her four pre-teen children have often witnessed their mother being verbally abused and physically assaulted by their father. The family is in dire financial straits, having little in the way of food and with outstanding electricity and telephone accounts, as well as being in arrears with the property rental. It is all too obvious that the woman is at her wit's end. The children appear withdrawn and listless, and have been infrequent in their school attendance. During the course of the visit the social worker observes two dogs, both restrained on short chains in the backyard, without adequate shelter or shade, with no clean water, and surrounded by mounds of their own excrement. Both dogs are in poor physical condition, being obviously malnourished and flea-ridden.

Another social worker is engaged by a child welfare agency to work with a 13-year-old male child who has been the victim of physical and sexual abuse. Over the course of many months the social worker observes that the boy has had any number of companion animals (birds, guinea pigs, kittens, puppies and rabbits) who have either disappeared without explanation or died with varying injuries for any number of vague reasons.

A fourth-year social work student, on her final placement, works with a family who have three young children, and who are experiencing myriad difficulties. The parents appear to be barely coping. At the commencement of her visits the social work student notices that the family has three puppies, and that the children regularly drop them or otherwise hold and carry them about in inappropriate ways. On subsequent visits she notices that the number of pups decreases from three to two, and then to one. Whilst she feels a degree of disquiet about their treatment and disappearance, she does not inquire as to their absence. The remaining pup looks obviously ill, but apart from suggesting to the children the correct way of holding the animal, the social work student, in spite of her concern about the animal's welfare and well-being, takes the matter no further. Upon further reflection, she then resolves that on her next visit she will voice her concern, but when she asks after the pup she is informed that the animal has died. The social work student feels guilt about not acting on her concerns, concerns that had progressively escalated during the course of her contact with the family, and agonises over how she might have done things differently.

A community-based social worker working with an elderly woman, without any immediate family and whose health has deteriorated rapidly over the course of the previous year, is faced with the dilemma that the woman's imminent admission to a nursing home means that her 15-year-old dog has either to be found another home or possibly euthanised. Either prospect horrifies the woman, and compounds her escalating anxiety at having to leave her own home.

In the final snapshot, a social worker is attempting to facilitate arrangements for a woman who is emotionally and psychologically fragile and is leaving her home and community due to longstanding domestic violence. The woman has made the decision to leave her children with her husband, but has a number of animals (a parrot, a cat and a horse) and insists that unless she is able to take them with her, in spite of everything, she will remain in her current unsatisfactory situation.

The importance of this book lies in its examination of the rationale for inclusion in, or exclusion from, social work's moral universe, rather than the discipline's current assumption that anthropocentrism is a valid and non-negotiable given. As part of the process of examining the supposedly obvious, it is essential to identify the philosophical and metaphysical assumptions that underlie the worldview of the Western tradition and, by association, the moral framework of contemporary social work, and to ask ourselves whether or not moral considerability is conveniently and irrevocably circumscribed by membership of the human species. Social work has unreservedly embraced this *a priori* assumption.

That this was not always so, as shall be shown later in this chapter, will undoubtedly be a source of great bewilderment given the incredulity with which one is routinely met at the mere suggestion that social workers give *consideration* to the question as to whether or not we should attend to the interests and well-being of animals, let alone commit ourselves to respecting subjectivity irrespective of species membership. This response is reflective of Midgley's (1996a, p. 232) observation that 'a tradition that a certain topic does not matter is one of the hardest things to get rid of – discussion tends to be tabooed before it starts'.

In the author's personal experience, a social work academic ventured the opinion that concern for animals had as much relevance to social work as did aeronautical engineering, another senior academic expressed misgivings that moral consideration of animals waylaid social work's attention from more *deserving* and *relevant* issues, whilst a social work practitioner (and Masters student) opined that it was *obvious* that domesticated animals are here but for us to eat, incredulous that their existence could possibly have any other purpose (Gompertz, 1992).

The negative appears so obvious that it is not even considered necessary to be dignified with, or justified by, reasoned moral argument; the well-being of animals and the natural world, apart from serving instrumental or aesthetic ends, is of no consequence or import for social work. Or so the orthodox refrain would have one believe. The notion that humankind is central to the cosmos is held as an article of faith, encapsulated in the belief that the world was either made *for* us, or is made *by* us, never leading us to pause for one moment to consider that maybe we are made *for* the world, and that 'our true good lies in enjoying and caring for the good in all things' (Clark, 1997, p. 96).

All five aforementioned scenarios are drawn from an amalgam of casework experience, and are presented by way of exploring what it is that social workers attend to and prioritise, and what it is that they ignore or relegate to relative unimportance. Even where, as in the case of the social work student, concern for the animals is all too obviously *felt*, and deemed to be of *some* importance, there appears to be no guide or source of intra-discipline illumination to which the social worker can refer in order to make an appropriate response, or engage in appropriate moral reflection and judgement. Paradoxically, social work folklore is replete with stories of animal neglect and/or abuse; it is often implicit in much casework experience, but infrequently made explicit. It is not as though social workers are unaware of, routinely indifferent to or negligent by malevolent intent of animal neglect and abuse. But it is also fair to say that social workers, by and large, rarely, if ever, act upon incidents that patently impact upon the welfare and well-being of domestic animals, let alone those animals routinely utilised in commercial agriculture, experimentation or any other number of activities where human interests are deemed automatically to trump those of our fellow sentient creatures.

In all of the foregoing scenarios, each social worker is compelled to wrestle with the selfsame question – who (in traditional language) is the 'client', and to whom, and to what, should they attend? In all probability each of the social workers would feel varying degrees of unease about the conditions and fate of the animals involved, but this may be not so much for the experience of the animals *per se*, but for what it might bode or entail for the character, or fate, of the humans involved:

> But still those innocents are thralls
> To throbless hearts, near, far, that hear no calls
> Of honour towards their too-dependent frail...

> (Hardy, 1976, p. 822)

Concern for animals and their well-being is marked by great ambiguity in Western societies, alternating between compassion and instrumentalism (Linzey and Cohn-Sherbok, 1997). Most prohibitions against cruelty are concerned with those animals with whom we share our lives and homes in a form of extended kinship, whilst the greater community of animals utilised for myriad human purposes is routinely subjected to treatment that would result in prosecution for cruelty or neglect were it occasioned in the domestic sphere. And yet we all too often betray the trust of even domestic animals; Coetzee's (2000, p. 78) Lucy pointedly observes that dogs 'do us the honour of treating us like gods, and we respond by treating them like things'.

Contemporary social work's moral framework does not allow for the moral considerability of any creatures *but* human beings, who are seen as not only possessing inherent value and dignity, but are deemed to be the *only* beings having such status. It is little wonder then that social work practitioners, their own *feelings* aside, routinely fail to act upon the neglect and abuse of animals, or take into consideration the importance of animals as extended family. The role that codes of ethics play in the failure of social workers to respond appropriately is a central moral concern of this book.

Even though it is well known in social work circles that perpetrators of domestic violence and sexual abuse/assault often injure or kill family animals (or at the very least vow to make good such threats), social workers by and large remain resolutely indifferent, even dismissive of according moral considerability to the aforesaid animals. A laudable initiative of domestic violence and sexual assault services in Launceston, Tasmania in the late 1990s to provide financial assistance for crisis shelter for animals of those accessing their services met with an initial degree of worker resistance and suspicion, on the grounds that funds should be employed to assist human victims and survivors, and this somehow detracts from or diminishes the experiences of those human beings. The substantial linkage between domestic violence, sexual assault, child abuse and animal abuse (Ascione and Arkow, 1999; Linzey, 2009a; Lockwood and Ascione, 1998) has as yet received negligible assent in social work circles, and compassion continues to be conceptualised as a rare and irreplaceable commodity, not, as Midgley (1983a) contends, a power of mind that increases with usage. When and where it is taken into account, negligible or nil attention is paid to the interests of the animals involved; consideration *has* to be presented in terms of human interests alone.

The omission of any individual, group or issue sends a loud and incontrovertible message that any and all of the aforementioned *do not matter*, that their interests are trivial, and that we ought to concern ourselves with more pressing issues. Social work's dogmatic anthropocentrism is metaphysical, conceptualising ourselves as different in *kind* from all other animals, and it serves to obscure our understanding of the human animal. It is assumed, not argued, that human beings are the measure of all things. And it is this absence that has particular implications for our consideration of the interests of animals. These are exactly the practical implications that routinely result from social work's moral indifference to the plight of animals; in practice, such moral dilemmas invariably tend to be resolved on species grounds. The 'client' is *always*, and *only*, the human being.

This book contends that social work's worldview, irrespective of theoretical orientation, is uniformly anthropocentric, and conspicuous exceptions to this orthodoxy are accorded scant attention in the mainstream, to be found (sleuth-like expertise is required to unearth them) as contributions to interdisciplinary studies (Loar, 1999; Loar and White, 1998; Quackenbush, 1981; Risley-Curtiss, 2009), a one-off limited format social work branch newsletter (noteworthy neverthe-less) dedicated to an exploration of the bond between human beings and animals (AASW, 1999a), a social work honours thesis given over to an exploration of why social work ought to concern itself with animal rights (Ryan, 1993a), and a subsequent journal article (Ryan, 1993b).

And one would look in vain for any guidance in coursework liter-ature specifically devoted to ethical deliberation (Banks, 1995; Levy, 1976; Lowenberg and Dolgoff, 1996; Reamer, 1995), or, for that mat-ter, in any Australian, British or American social work code of ethics. Animals are conspicuous by their absence, and the scope of social work ethics is invariably treated as inherently anthropocentric. When curso-rily considered, the issue is either dismissed on the grounds that to do so would lead us well beyond the confines of our Western tradition (Clark and Asquith, 1985), or even where consideration is acknowledged (even these exceptions devote no more than a couple of pages, at most), nor-mative implications are not explored because the subject matter is held to be beyond social work's legitimate ambit (Clark, 2000; Downie and Telfer, 1969, 1980; Ife, 2001). These oddities aside, the invisibility of animals in social work literature merely serves to confirm the percep-tion that expressions of concern for animals are at best misplaced, and at worst frivolous, preposterous and misanthropic to boot.

To understand how it is that social work thinks in this way and where we go from here (to provide practical guidance to its practitioners), the Western worldview will be briefly examined, particularly its meta-physical and philosophical conceptualisations of the moral standing of animals, in order to provide the context within which social work operates. Social work's criteria for moral regard and considerability are thoroughly human-centred, and this singular failure to acknowledge our embeddedness and biological continuity, and its moral implications, as well as its uncritical acceptance of orthodoxy, are the central problems that will occupy this work.

Worldviews, or our *imaginative visions* (Midgley, 2001), have profound implications for how we define ourselves, as well as for our understand-ing of our place in the natural world and our responsibilities to other species. There exists an important correlation between morals and meta-physics (Murdoch, 1993), for 'The ways in which we imagine the world determines what we think important in it' (Midgley, 2003, p. 2). In the Western world there are enduring metaphysical, philosophical and the-ological traditions that view humankind as not only unique, but whose dignity demands a transcendence of animality and the natural world; Schweitzer (1955, pp. 228–9) comments that

European thinkers watch carefully that no animals run about in the fields of their ethics...Either they leave out altogether all sympathy for animals, or they take care that it shrinks to a mere afterthought, which means nothing. If they admit anything more than that, they think themselves obliged to produce elaborate justifications, or even excuses, for so doing.

Orthodoxy, Fuller (1949, p. 833) observes, supposes that 'the only way of keeping the system in order and man master of it is to shoo them out of the house altogether and stop one's ears against their scratchings at the door'. Indeed, 'The love of animals is often spoken of by intellectuals as an example of modern sentimentality' (Clark, Kenneth, 1977, p. 45).

In all cultures animals assist in the definition of what it is to be human (Baker, 1993; Kalof, 2007; Thomas, 1983), and Salisbury (1994, p. 11) observes that 'Our attitudes towards animals, our treatment of animals reveal our attitudes towards ourselves.' Our similarities and dissimilarities have preoccupied humans across the ages, and Thomas (1983, p. 40) observes that 'Neither the same as humans, nor wholly dissimilar, the animals offered an almost inexhaustible fund of symbolic meaning.' Their very being is utilised as an affirmation of our *otherness*, and entails their disparagement as embodiments of our all too human failings, leading to what Mason (1993) terms *misothery*, hateful and contemptuous attitudes towards them.

Whilst the Judaeo-Christian tradition is often deemed to be responsible for this radical disjuncture, and to be inherently antithetical to concern for animals and the natural world (Singer, 1976; Toynbee and Ikeda, 1976; White, 1967), this is by no means held uniformly to be the case (Clark, 1993; Linzey, 1987; Linzey and Yamamoto, 1998). Indeed Preece (2005) argues that Christianity has engendered more respectful attitudes to animals than has Darwinism. Nevertheless, one need not look far for evidence of neglect – the author of *Genesis* (9: 2–3) decrees that 'The fear of you and the dread of you shall be upon every beast of the earth...air...sea...into your hand are they delivered', interpreted by Aquinas as sanctioning *absolute* dominion (Aquinas, 1989). Augustine (1990) holds that human rationality mirrors the divine, and because animals lack reason we have no need to concern ourselves with their suffering, for their purpose is human ends.

Robert Burns (Wynne-Tyson, 1985, p. 40) laments

> I'm truly sorry man's dominion
> Has broken Nature's social union.

For all its obvious shortcomings the Judaeo-Christian tradition ultimately held human beings accountable for their actions to a being
greater than their individual and collective selves (God's existence
reminds us that we are *not* God), and in theory, at least, conceptualised
human beings as stewards of creation (Linzey, 1987; Linzey and Cohn-
Sherbok, 1997), a notion that has evolutionary plausibility (Midgley,
1983a). Furthermore the Biblical notion of a covenant between God
and all living things entails the rejection of an exclusive human community in preference to a mixed community inclusive of animals, and
finds parallels in the metaphysics of the Eastern traditions of Buddhism,
Hinduism and Jainism (Chapple, 1993).

The prophet Nathan (*II Samuel* 12: 3) reflects a quite different relationship to that conceived to be the norm – 'the poor man had only
one little ewe lamb that he had bought. He tended it and it grew up
together with him and his children: it used to share his morsel of bread,
drink from his cup, and nestle in his bosom; *it was like a daughter to him*'
[emphasis mine]. Reflecting on the moral of this tale, Midgley (1983a)
notes that the poor man's solicitude is no bourgeois folly, it is not a substitute for absent children, and his affection and relationship with the
ewe lamb is perfectly natural. In other words, it is appropriate that we
exhibit loving-kindness and affection towards humans and animals, for
their own sakes.

Freed from the constraints of spiritual traditions, the natural world
was duly divested of associated feelings of awe and wonder (Sheldrake,
1991), and Thomas (1983) claims that the basis for the contemporary
concept of balance in nature was theological before it was scientific. Within the Christian tradition St. Francis of Assisi proposes a
radically compassionate conception of our relationship with animals,
one that acknowledges ontological continuity and our shared origins
(Armstrong, 1973; Sorrell, 1988). We find parallel sentiments in the
world of literature, where Dostoevsky's (1952, p. 167) Father Zossima
entreats us to 'Love all God's creation... Love the animals, love the
plants, love everything. If you love everything, you will perceive the
divine mystery in things... And you will come at last to love the whole
world with an all embracing love.'

By way of contrast, Callicott (1982) identifies Western natural philosophy, with its much greater influence upon modern scientific thought,
as the chief culprit. Modern materialism, with its roots in the reductive
and mechanical ontology of the Greek atomists, entails a monadic moral
philosophy that posits two fundamental options for ethics – either a

Hobbesian harmonisation of our appetites, or a conceptual talisman like Kant's Reason (Callicott, 1989).

Whilst Greek and Roman philosophers did not employ the language of rights, Nash (1990) observes that their utilisation of the principles *jus naturae* were a recognition that prior to human-made law human beings lived according to certain biological principles, which were in due course overlaid by conceptions of justice, *jus commune* or the common law. The Greek and Roman philosophers were also greatly preoccupied as to the nature of the relationship between human beings and animals. As a consequence the Romans articulated the moral precepts *jus animalium*, which inferred that animals, independent of humankind and their legal and political structures, possessed what subsequent philosophers referred to as *inherent* or *natural rights*, and Nash (1990, p. 17) observes that 'As the third-century Roman jurist Ulpian understood it, the *jus animalium* was part of the *jus naturale* because the latter includes "that which nature has taught all animals; this law indeed is not peculiar to the human race, but belongs to all animals".'

The Magna Carta of 1215, which ushered in the concept that eighteenth-century revolutionaries termed natural rights, is characterised by Nash (1990, p. 13) as 'ethical dynamite', observing that this concept possesses a tendency to take on expanded meaning and has within it the seeds of revolutionary ethical extension. Its significance lay in the attribution of rights by mere virtue of existence, albeit limited to a specific section of society, the English male nobility, and by placing limitations upon the exercising of royal power. The natural rights due to human beings issue from the *nature* of human beings, and these moral entitlements are pre-conventional in that they are not derivative of any legal, moral or contractual criteria (Rowlands, 1998). Natural rights principles took root and bore fruit in Western thought throughout the seventeenth and eighteenth centuries, and Thomas (1983) argues that there were three forces at play in bringing animals into the circle of moral considerability – firstly, the admittedly minority Christian position that humans were God's stewards; secondly, the awareness that the world had value apart from human ends, which, he claims, represented a revolutionary development in Western thought; and thirdly, sentience, rather than rationality, was accorded *the* criterion for moral standing.

Not unexpectedly, the Western philosophical tradition is a mixed bag; Dombrowski (2000) claims that whereas the Stoics were unambiguously dismissive of animals, Plato and Aristotle were more respectful (if ambivalent), whilst Plotinus, Plutarch, Porphyry and Pythagoras were part of the tradition that emphasised similarities between species, a sense of kinship and responsibilities to other animals (Preece, 2008). Aristotle (1952) conceptualises the natural world as hierarchical in nature, with the role of the less rational and imperfect beings to serve the interests of the more rational and perfect, and Steiner (1998, p. 272) contends that these two beliefs 'are foundational for the entire subsequent tradition of Western thinking regarding the relationship between human beings and animals'. Even so, Aristotelian ethics (that which is requisite for the flourishing of all animals) are informed by biology (Aristotle, 1952; Clark, 1975), but Clark (1999) argues that Plato, received knowledge notwithstanding, posits a greater ontological continuity, and duties to animals not of our kind, than does Aristotle; indeed, Aristotle maintains that our responsibilities to animals are lessened by their inability to share a sense of community with humans.

Cartesian, Hobbesian, Spinozian and Kantian formulations of ethics are intrinsically human-centred, with subjectivity deemed to be incontrovertibly the sole preserve animals of *our* kind. Human beings, Hobbes (1904) argues, are calculating egoists, and our need of others is predicated solely on their utility to us; animals, having no language, and therefore unable to enter into contracts, fall outside the circle of moral standing. Descartes (1989) avers that animals cannot be harmed or suffer because they are insensible and irrational machines, for reason is dependent upon speech.

Descartes famously declared '*Cogito, ergo sum*' [I think, therefore I am], thereby installing rationality as absolute sovereign. It was Descartes' metaphysics that paved the way for a radical rendering asunder of human beings from the natural world, and inculcated a reductionist and instrumental view of all fellow animals; Marx (1990) likewise conceives consciousness as an exclusively human capacity. By way of contrast, Spinoza (n.d.) insists that ethical concern for others is dependent upon their being like us, and whilst conceding that animals feel, their exploitation receives its justification from their essential dissimilarity to human beings.

Kant (1990) holds that rationality, not consciousness, is *the* characteristic that qualifies a being for moral regard, and as animals are not *self*-conscious – he nevertheless credits animals with an emotional life (Midgley, 1996a) – we have no direct duties towards them, for they exist merely as means to human ends. Central to Kantian philosophy is the distinction between persons and things, between subjects and objects, and as important as this distinction is for how we view and treat human beings, it specifically excludes animals from the moral universe given that only rational beings can be persons and subjects.

Likewise, Aquinas dismisses animals as deserving of our charity due to their being irrational creatures, and Kantians and Thomists alike view duties to animals as merely indirect duties to our fellow humans, their value contingent upon our interests, 'to be cared for only as a practice for caring for people' (Clark, 1997, p. 47).

But this was not the uniform philosophical position; indeed there has been an historical discrepancy between the proclamations of metaphysicians and philosophers and everyday thinking about animals, and there were discordant, albeit minority, voices raised against the blanket exclusion of animals from the circle of moral considerability (Harwood, 2002; Sorabji, 1993; Steiner, 2005).

By way of example, Rousseau (Donovan, 1996), whilst maintaining that animals are neither rational nor autonomous beings, nevertheless holds that they are indubitably *sentient* creatures. The view that sentience, rather than rationality or free will, ought to be the decisive attribute that grounds our moral duties towards animals, finds its parallel in Bentham's (Singer, 1976, p. 8) conviction that 'The day *may* come when the rest of the animal creation may acquire those rights which could never have been withholden from them but by the hand of tyranny...The question is not, Can they *reason*? nor Can they *talk*? but, *Can they suffer?*' It is Donovan's (1996) belief that Rousseau and Bentham consider sentience, a common condition that unites human beings and animals, to be *the* attribute that secures entrance into Kant's kingdom of ends.

The notion that moral standing be confined to rational beings and that all others be uniformly consigned instrumental status is the cause of great indignation in Schopenhauer (Midgley, 1983a, p. 52), who insists that 'Boundless compassion for all living beings is the firmest and surest guarantee of pure moral conduct, and needs no casuistry...If we attempt to say, "This man is virtuous, but knows no compassion"...the contradiction is obvious.'

What this brief overview evidences is the prevailing tendency in the Western tradition to equate moral status and subjectivity with rationality. That which orthodox Christianity promulgated was in turn appropriated by humanism, with the worship of God being replaced by the worship of humanity. Whereas animals had been seen as God's creatures and the natural world a manifestation of God's beneficence and grandeur, humanism declared that humankind alone had inherent value. In the Middle Ages animals were held accountable for myriad crimes against human beings and their property, and were subjected to both legal and ecclesiastical jurisdiction, resulting in criminal prosecution and capital punishment, as well as exorcisms and excommunications (Evans, 1906; Hyde, 1915/1916). As bizarre as this might sound to modern ears, such practices at least accorded to animals a modicum of justice, and bestowed upon them a status of being viewed as more than mere objects.

Humanism, in trumpeting humankind as the *summum bonum* of creation, was more dogmatically anthropocentric than the religious traditions it sought to supplant (Clark, 1997). What humanist moral thinkers feared above all else was that human beings would assuredly be treated like animals should the metaphysical and moral divide between human beings and animals be, in any significant manner, breached. Human dignity required a transcendence of the natural order, and comparisons with animals were considered an anathema. Whereas the religious Everyman had been shackled by the dictates of blind faith, all animals were creatures enslaved to blind instinct. The metaphysical map of humanism contrasted rational (and hence moral) Man with irrational (and therefore non-moral) animals, a world neatly divided between subjects and objects, persons and things. Humanity became self-absorbed and partial, content to gaze and marvel at its own reflection like a modern-day Narcissus.

The most profound challenge to both religious and humanist worldviews, and human hubris, came with the publication of Charles Darwin's *On the Origin of Species* in 1859, though it needs noting that evolutionary theories had well and truly preceded Darwin (Elsdon-Baker, 2009). From our contemporary vantage point it is difficult to imagine, let alone appreciate, the degree of controversy that Darwin's theory of evolution engendered, with Ryder (1983, p. 132) likening its impact to

> a bombshell which blasted man's arrogant assumption that he was in a superior and separate category to all other animals. Once man had reluctantly accepted that he was just one species among many others...and indeed shared kinship with them...He could no longer justify an entirely separate moral status for himself. If there was biological kinship then why not a moral kinship also?

Thomas Hardy (1976, pp. 906–7) captures it thus:

> Next this strange message Darwin brings,
> (Though saying his say
> In a quiet way);
> We all are one with creeping things;
> And apes and men
> Blood-brethren,
> And likewise reptile forms with stings.

Darwin's worldview, which would forever alter the way in which we view ourselves and our place in the natural world, refutes the notion of an ontological divide between ourselves and all other species, and furthermore contends that morality itself developed out of the process of evolution. Human beings and animals, Darwin (1936, 1965) insists, share a common ancestry, with the difference in mental and emotional capacities being one of degree, not kind. Darwin (1936, p. 494) is of the considered opinion that as ethics progressively evolve *all* sentient beings will come to be included within the sphere of moral concern, resulting in a 'disinterested love for all living creatures'.

It is to be noted however that Darwin's espousal of a moral egalitarianism was to have minimal effect then *and* now. Paradoxically, the Darwinian revolution, which pointed us in the direction of our ontological continuity, with its consequential moral implications, was interpreted almost invariably to reinforce anthropocentrism. Rather than encouraging fellow feeling and compassion, Clark (1977, p. 30) laments that 'Unfortunately, the symbolic use to which Darwinism has been put is to exalt Man as heir of the ages, and depress the non-human animals as errors, or backslidings, or material', and Gray (2002, p. 31) opines that 'Darwinism is now the central prop of the humanist faith that we can transcend our animal natures and rule the earth.' Such interpretations have impeded serious moral consideration of the nature of the relationship between ourselves and all other animals (not to mention the understanding of ourselves), as well as the nature of our duties and obligations to them.

Social work's innate suspicion of biological considerations and explanations, and (it will argued) its consequential dismissal of any moral consideration of animals, is in large part due to the deleterious moral and ethical implications of Social Darwinism, and contemporary ultra-Darwinism. As a consequence, Darwinian theory itself was rejected out of hand for its seemingly misanthropic intent and implications. That this was not a uniform response by early social workers to Darwin's theory is borne out by Charles Brace, a leading American social worker and reformer, who discerned in evolution the basis and efficacy of human virtue and the assurance of human perfectibility (Hofstadter, 1965).

Precisely because Social Darwinism was so strident in deducing egoistic ethical precepts from Darwinian theory, it was invariably assumed that the only possible Darwinian approach to ethics was an egoistic rendering (Midgley, 1995a; Williams, 1974). Such an assumption was to have significant implications for how the developing discipline of social work would conceptualise human nature and our place within the natural world, as well as social work's conceptualisation of morality. In a reaction to what it perceived to be the deterministic and fatalistic implications of Darwinian theory, social work conceived human nature as interminably malleable, and so commenced social work's extolling of nurture to the virtual exclusion of nature. The tragedy lies not in its rejection of Social Darwinism, rather in its rejection of Darwinian theory *per se*.

The earliest legislation anywhere in the world specifically relating to animals was 'An Act to Prevent the Cruel and Improper Treatment of Cattle', also known as Martin's Act, in Great Britain in 1822, which made cruelty a punishable nationwide offence. Martin's Act was phrased in terms of injury to *other* persons' animals and therein reflected the disinclination of those drafting legislation to address cruelty inflicted upon animals by their owners. Nash (1990) draws parallels with this aversion to intervention with the disinclination of contemporary law drafters to respond to early attempts to ameliorate master–slave and husband–wife relationships. Likewise, concern for animal welfare was disproportionately centred upon the pursuits and interests of the working class (Harrison, 1967; Lansbury, 1985; Turner, 1980), and the state intervened on behalf of domestic animals decades prior to doing so on behalf of women and children (Kean, 1998). Those humanitarian reformers who sought to draw attention to the inherent linkage between human and non-human welfare were condemnatory of such obvious contradictions (Ryder, 1989; Salt, 1980).

The Society for the Prevention of Cruelty to Animals, later to become the Royal Society for the Prevention of Cruelty to Animals (RSPCA), was established in 1824 by Anglican clergyman Arthur Broome and fellow humanitarians, several of whom (including Richard Martin, Thomas Foxwell Buxton and William Wilberforce) had also been prominent advocates for the abolition of slavery, workplace and social reform, as well as child protection (Moss, 1961; Ryder, 1989). This organisation played a pivotal role in the promotion of legal reform, including the outlawing in 1835 of fighting or baiting any badger, bear, bull, cock, dog or any other animal; the prohibition of the use of carts drawn by dogs in 1854; the regulation of vivisection in 1876; and the regulation of animals in transit in 1878 (Ryder, 1989). Ryder notes, by way of comparison, that the dissection of living animals in the United States went unchallenged and unrestricted until the Animal Welfare Act of 1966.

In its formative stages the animal welfare movement was fostered by Christian moralists, amongst whose numbers were Evangelicals, Methodists and Quakers, as well as sceptics (Attfield, 1983; Li, 2000). Nineteenth-century humanitarians conceived the interrelatedness of cruelty, suffering and oppression as transcending species boundaries, and their conception derived from natural rights principles that averred that oppression and suffering were not peculiarly human experiences. Compassion and empathy were not conceptualised as rare and irreplaceable resources, rather a habit or power of mind (Midgley, 1983a), a viewpoint that was accorded legitimacy by the humanitarian reformers who spoke out as vociferously in defence of human beings as for animals. Martin was also involved in the promotion of laws that sought to guarantee protection to child labourers in 1802 and 1833 in Great Britain. Shaftesbury and Wilberforce, the most historically lauded social reformers of the nineteenth century, conspicuously devoted the latter years of their lives in engaged in advocating on behalf of the welfare of animals. Shaftesbury, who steadfastly opposed the practice of vivisection, made significant contributions to reforming legislation that dealt with the treatment of the insane, the employment of children as chimney sweeps and in mines, the condition of workers in mills and factories, as well as measures designed to ameliorate the education and the housing of the poor (Ryder, 1989).

Simultaneous concern for the well-being of human beings (*especially* those formerly despised) and animals gathered significant momentum in the nineteenth century, reflecting a widening of moral sensibilities (Harrison, 1982; Weinbren, 1994), and political and social change (Gould, 1988; Kean, 1998). There exists, Salt (1980) insists, an intrinsic relationship between oppression and cruelty and a dearth of imaginative sympathy and kinship; sentience and individuality in *all* animals are what matters morally, with the essential prerequisite for change being that oppressors come to recognise those they oppress as being members of their own community. Historically, the utilitarian tradition sought to expand the moral circle beyond human interests, and Mill (1901, p. 17), following in the footsteps of Bentham, argues that morality and membership of the moral community is to be 'secured to all mankind; and not to them only, but so far as the nature of things admits, to the whole sentient creation'. Utilitarianism holds that 'the point of morality is ... the happiness of beings in *this* world' (Rachels, 1995, p. 92).

It was not only in the British Isles that the pioneers of legal action against child abuse happened also to be advocates for the welfare of animals. In 1866 Henry Bergh, the founder of the American Society for the Prevention of Cruelty to Animals, with the assistance of ASPCA attorney Elbridge Gerry and Quaker activist John Wright, was instrumental in establishing the Society for the Prevention of Cruelty to Children from a corner in the ASPCA office (Clifton, 1991). Legislation specifically enacted to combat cruelty to animals was utilised to counteract ill-treatment of children, with Bergh (Scott and Swain, 2002, p. 7) averring that 'The child is an animal ... If there is no justice for it as a human being it shall at least have the rights of a stray cur in the streets. It shall not be abused.'

In Great Britain the RSPCA established the National Society for the Prevention of Cruelty to Children (NSPCC) in the same year, and the endeavours of Shaftesbury, Cardinal Manning and Dr. Barnardo were instrumental in its creation (Turner, 1992). The commonly acknowledged founder of the NSPCC, the Reverend Benjamin Waugh, professed the Society's debt to the RSPCA by expressing that 'Your Society, the RSPCA, has given birth to a kindred institution whose object is the protection of defenceless children' (Ryder, 1983, p. 132); as Lansbury (1985) observes, attention to the well-being and welfare of children was a natural outgrowth from human compassion towards animals. Parallel trends were likewise evident in the Antipodes; the Queensland Society for the Prevention of Cruelty was initially established in 1883 specifically for the protection of animals prior to turning its attentions to the plight of children, whilst the Victorian Society for the Prevention of Cruelty to Animals extended its brief in late 1884 so as to include concern for children (Scott and Swain, 2002).

The American Humane Society (AHS), founded in 1877 with the express objective of achieving federal laws to ensure the protection of animals involved in interstate commerce, expanded to become a society for the protection of animals *and* children in 1885. Of particular interest is the fact that the American Humane Society was subsequently involved in the creation of the International Children's Congress in 1913, and the International Child Welfare League in 1916. The AHS was intimately involved with myriad human welfare issues, including the promotion of legal rights and humane care for orphans and illegitimate children; the combating of baby-selling rackets; the initiation of road safety education for children in 1926; and as recently as 1948 AHS affiliates were engaged in the provision of shelter to 50,000 children from abusive situations every year. Contemporarily, the AHS lobbies for legislation that promotes the welfare of children and is best known for its studies and accompanying statistical data relating to myriad child welfare issues (Clifton, 1991). The close connections between animal and child protection societies in both Great Britain and the United States belie the contemporary notion that human and animal concerns are unrelated, and is a classic example of the concomitancy of 'the association of humanity to animals with humanity to humans' (Ryder, 1983, p. 132).

In late Victorian and Edwardian Great Britain, campaigns on behalf of animals were increasingly associated with crusades for the advancement of the interests of women, and significantly the majority of positions of leadership in animal welfare societies were occupied by women, who were also disproportionately represented in the movement as a whole (Elston, 1990; Leneman, 1997). Indeed French (1975) notes that female participation in the anti-vivisectionist movement was amongst the highest of any cause without explicitly feminist objectives. The most formidable anti-vivisectionist of this period was Anglo-Irish writer, feminist, journalist and social worker Frances Power Cobbe, who in her prolific literary outpouring succinctly clarified the linkage between anti-vivisection, philanthropy and the women's movement (Elston, 1990). In the 1850s, whilst in her early thirties, Cobbe assisted a pioneering social worker, Mary Carpenter in the teaching of children in the Bristol slums (Ryder, 1989). Cobbe held cruelty to constitute the worst evil, and devoted substantial time and energy to the cause of animal welfare, but the rights of human beings, especially those of women, remained her primary concern. Cobbe campaigned for the legal protection of women and the reform of the institution of marriage, as well as advocating for university places and women's suffrage. Paralleling Cobbe in the United States was the prominent animal welfare advocate Caroline Earle White, who had initially been passionately engaged in the movement to abolish slavery, but who was to spend the remainder of her life working on behalf of women, children and animals (Clifton, 1991).

Cobbe's anti-vivisection campaigns had the subsidiary goal of facilitating the improvement of women's economic, political and social positions; her employed tactic of bestowing feminism with a mantle of respectability imbued with moral conservatism greatly enhanced the prospects of both causes (Elston, 1990). Other prominent feminists who were also advocates for the welfare of animals were Annie Besant, Charlotte Despard, Anna Kingsford, Lewis Gompertz, John Stuart Mill and the social worker Maria Dickin, who founded the People's Dispensary for Sick Animals of the Poor (PDSA) in 1917 in London, and whose work was subsequently extended nationwide. Dickin, who initially worked amongst the poor in London's East End slums, came to be increasingly horrified at the conditions and welfare of animals with which she was regularly confronted, and believed that efforts to create a better world must of necessity include our fellow animals (Dickin, 1950). Her dispensaries provided free veterinary treatment for sick and injured animals of the poor, and in time she established an Animals' Sanitorium and animal hospitals. Dickin's work was transferred by her workers in subsequent years to other European nations as well as to North Africa, and the PDSA continues to this day to provide the practical services established by Dickin (PDSA, 2010).

Following Cobbe's departure, opponents of the movement to enhance the welfare of animals, and their moral standing, played upon male prejudice by derogatively equating concern for animals as exhibiting 'womanly' tendencies, inferring sentimentality, irrationality, lack of objectivity, hysteria and squeamishness (Clark, 1997; Midgley, 1983a). Such an attribution was utilised to ridicule the validity of human concern for animals, and was exceedingly effective in marginalising and rendering invisible both the issue and the advocates themselves. Given that the overwhelmingly majority of early social work practitioners were women, the fledgling discipline was itself no stranger to the role that sexual politics played in the wider society and in marginalisation. This process had far-reaching ramifications, and in particular it ensured that for the majority of the twentieth-century feminism would fail to make linkage between the oppression of women and animals, although this omission is now being rectified (Adams and Donovan, 1995; Birke, 1994; Donovan and Adams, 1996).

Even the passion of the aforementioned individuals cannot mask the reality that they were essentially minority voices always struggling merely to be heard, let alone be accorded any modicum of seriousness. Since the early twentieth century, social work has assiduously excluded animals from both its moral map and sphere of interest. Contemporary social workers would quite understandably assume that this has always been the case and, in keeping with tradition, quite likely assume that thus it ever will be. One can safely venture that they would be quite taken aback to discover that in the nineteenth century there existed clear linkage between human and animal concerns, and that early social workers accorded explicit attention to the well-being of both groups. The reason or reasons for the discipline's contemporary indifference would have likewise been a source of consternation to many social work pioneers. It is a matter of speculation as to why this state of affairs came to pass, and to be seen as not only obvious *but* essential.

The fact that concern for animals in the nineteenth and early twentieth centuries focused overwhelmingly upon members of the working classes is surely a not-insignificant factor then *and* now for social work's dismissal of moral consideration of animals, given that social workers have historically worked almost exclusively with the weak and vulnerable in human society (as witnessed in the practice scenarios). It also needs saying that whilst the influence of sexual politics was not insignificant, these were not the only, or indeed preeminent, reasons for the marginalisation of concern for animals. The dual influences of psychiatry and sociology on the development of social work post-World War I served to focus nigh-exclusive attention on the human individual or human society in abstraction from the natural world, whilst social work's desire to be accorded professional status surely played a not insignificant role in the jettisoning of animals from the discipline's sphere of interest.

Allied to these reasons is the almost universal notion that social work values are by definition exclusively human-centred; no animals stray into social work's house, let alone have a place at morality's high table. Anthropocentric orthodoxy was and remains the chief villain of the piece – the exclusion of animals from the moral sphere was seen as obvious, and as proof of one's concern for human beings. Zoophilism and misanthropy were, and are, invariably taken to be natural bedfellows. But social workers cannot disregard moral arguments just because they have always been ignored thus far, 'because they prefer to think the present obvious. How things are is not how they must be' (Clark, 1991, p. 121).

Whilst it is fair to say that it is extremely unlikely that any contemporary social worker would advocate cruelty to animals, the majority would object to it on the grounds that it may well lead to cruelty to human beings, whilst many are seemingly indifferent even to this reality (given the almost deafening silence on the subject), and most, one suspects, are not particularly clear in their own minds what they think about this issue or why. To be sure, 'it does not make sense to morally care about the suffering inflicted upon other animals unless some value is placed upon the lives of those animals' (Godlovitch, Roslind, 1972, p. 171). All in all, it represents a most unsatisfactory state of affairs.

All that remains to be is to provide an outline of the focus of each subsequent chapter.

Chapter 2 will provide confirmation that social work is a discipline that is essentially concerned with respect for the subjectivity of human beings, and that social work has an inherently moral dimension. The chapter will also undertake an exploration of how moral inclusion or exclusion is dependent upon how social work thinks about itself, and by way of an examination of the concepts of moral considerability and the moral community, and reflection upon the origins of morality, an argument will be mounted that moral standing cannot be circumscribed by the species barrier.

Chapter 3 undertakes an extended exploration of the nature of human beings and animals, arguing that an acknowledgement and acceptance of our status as terrestrial creatures, allied with a greater understanding of and respect for our fellow animals, is perfectly consistent with human dignity and moral standing. Rejecting both biological and cultural determinism, it will be argued that our biological continuity and similitude with other species has moral implications.

Chapter 4 will undertake an examination of *the* cardinal value of social work and morality generally, that being the principle of *respect for persons*. It will argue that this principle, as traditionally conceptualised, is problematic for many human beings, and arbitrarily excludes our fellow animals from the moral community. A moral framework will be articulated that is inclusive of *all* animals, human and non-human.

Chapter 5 will provide some practical and theoretical implications for social work practice consistent with the adoption of the principle of *respect for individuals*, by means of reflection upon the casework scenarios outlined in Chapter 1. Rather than being definitively prescriptive, it is hoped that these reflections will serve to prompt further sustained moral reflection by social work students, practitioners and academics.

Finally, the Appendix will articulate a new social work code of ethics that is inclusive of animals, one that grounds a compassionate and respectful practice towards both humans and animals, for 'each creature is the outward sign of an equal soul, and to be respected as such' (Clark, 1984, p. 177).

# 2
# Social Work, Subjectivity and the Moral World

Our whole life is startlingly moral.

*Henry David Thoreau* (1968, p. 192)

Social work is a moral discipline, and the very existence of social work values presupposes a moral framework and a vision of a moral community. This chapter is foundational, for we need to understand why social work conceptualises morality and moral considerability in the way it does, and how this impacts upon our understanding of both humans *and* animals. The chapter will argue that the subjectivity of our fellow animals makes the widening of the moral circle, and thus the scope of a code of ethics, morally obligatory. It will also examine accounts of the origins and functions of morality.

Social work conceptualises human beings as social beings, and acknowledges the inherent tension between its being an instrument of social change and social stability, having duties and responsibilities to the individual and society. Notwithstanding its theoretical multitudinousness and contested definitions, social work is underpinned by shared values and moral framework, and can therefore be seen to be a definable, bounded discipline. Social work has historically attended to the human individual as the primary locus of moral value, and social work is best characterised by its moral attentiveness to subjectivity, which underpins and guides social work practice. That being so, this book is interested in understanding why is it that social work does not extend *any* moral consideration to the subjectivity of other creatures.

Contemporary social work tends to be dismissive of the notion that it has an explicit moral mission of assisting people to lead good lives in a good, and just, society (Siporin, 1982), and it's little wonder then that social work is increasingly straying from its core mission of serving

21

society's weak and vulnerable (Specht and Courtney, 1995). A partial explanation for this trend is the penchant for professional status, as distinct from a discipline motivated by specific moral principles, which currently permeates social work. Increasingly pursued as a career rather than a calling or vocation, it is no longer a rarity to come across social workers who practice perfunctorily, who are singularly indifferent to the indigent, as *officially* and ostensibly opposed to indigence. One is reminded of Gandhi's conviction that for the poor God always appears firstly in the form of food and shelter, and Chesterton's (Canovan, 1977, p. 36) observation that 'If we want to talk about poverty, we must talk about it as the hunger of a human being... not that there is insufficient housing accommodation, but that he has not where to lay his head.' Even (often especially) agencies trumpeting social justice seem oblivious to this reality, and it was for such reasons that Mother Teresa (Wilkes, 1985, p. 42) refused the title social worker, observing that 'social workers work for a cause; the Church works for people'.

References to 'the value base', 'the values underpinning practice' or that social work is a 'value-laden activity' are so ubiquitous that an onlooker would assume values to be the stock-in-trade of social work. However, apart from works specifically addressing ethics, one is struck by the paucity of treatment accorded to the role of values, let alone moral philosophy, in the broader social work literature. Discussion of values and morality tend to be conducted in a generalised, cursory and at times desultory manner, and this observation includes major theoretical texts (Howe, 1987; Payne, 1997; Turner, 1996a). Values are more often than not treated as though they are self-evident and self-explanatory entities, and the language of the literature dealing with values and morality tends to exhibit rather than analyse (Timms, 1983).

Social work values are largely derivative of the Western moral tradition (Clark and Asquith, 1985; Plant, 1970), and the traditional conception of human life as both unique and sacrosanct has provided the bedrock for the fundamental moral and social values of Western society (Wilkes, 1981). Kant posits that all moral thinking is predicated upon an awareness of otherness (Midgley, 1996a), whilst Todorov (2000) characterises all morality as concerned with an awareness of other individuals. The emphasis upon freedom, and recognition of the individual, were significant features of Enlightenment thought (Harmon, 1964), and brought forth the evolution of a theory of rights for human beings, preparing the way for truly revolutionary progress towards social justice (Midgley, 1996a). Its extremism, however, was manifested in its decree that human beings were taken to be the measure of *all* things, and that human dignity demanded an irrevocable divide between human beings and all other animals.

This chapter will be interested in discerning why it is that both values and morality are conceptualised as being limited in scope to human beings, and the consequences this has for our conceptualisation of moral value and considerability. It will be argued that the exclusion of animals from social work's moral universe is the consequence of ignorance and indifference rather than premeditated malign intent, for as evidenced in Chapter 1, this has not always been a uniform feature of social work.

That social work is an undertaking fundamentally concerned with issues of an ethical, moral nature is incontrovertibly confirmed by social work literature (Banks, 1995; Butrym, 1982; Clark, 2000; Clark and Asquith, 1985; Elliott, 1931; Horne, 1987; Imre, 1982; Levy, 1976; MacCunn, 1911), and it is argued that social work is shaped and informed to a far greater extent by morality and values than it is by theory. Whilst 'responsible, ethical practice needs to be built on strong theory' (Turner, 1996b, pp. 10, 12), it is nevertheless acknowledged that 'theories are value driven'. Social problems are essentially moral problems, and social policies differ markedly from economic ones, for their aim lies in creating 'moral relationships between individuals, giving individuals moral identities in relation to others' (Watson, 1980, p. 8). Any society bereft of common social ethics tends to substitute rights for functions, and self-interest for social purpose (Tawney, 1930).

The literature also shows that social work is fundamentally concerned with the human subjectivity (Biestek, 1973; Moffett, 1968; Perlman, 1979; Ragg, 1977; Richmond, 1922; Robinson, 1930; Wilkes, 1981; Woods and Hollis, 1990). Social work, it is said, 'emphatically embraces human subjectivity' (Pearson, 1975, p. 128), 'is concerned to produce a knowledge of man as a subject' (Philp, 1979, p. 91), and 'is essentially concerned with producing a knowledge of the individual as a subject' (Horne, 1987, p. 86).

Central to social work's self-definition is its conviction that human beings possess inherent and inalienable value; individuals, Biestek (1973, p. 25) insists, have a right 'to be treated not as *a*, but as *this* human being'. Such sentiments are ubiquitous in social work literature; Goldstein (1973, p. xiii) claims that 'distinct individuals' are the ultimate concern of social work, Hollis (1972, p. 14) regards it as *the* 'fundamental characteristic of casework', and Shaw (1974, p. xiii) claims that social work 'is about understanding the individual, rather than knowledge about people in general'.

Whilst social work knowledge has historically been primarily derived from the disciplines of psychology and sociology (Butrym, 1982; Wilkes, 1981), its philosophical roots can be traced to Christianity (Bowpitt, 1998; Niebuhr, 1932; Siporin, 1986) – Calvinism and Pauline theology (Grimm, 1970) in particular, and classical liberalism (Clark and Asquith, 1985; Woodroofe, 1971) – and the manner in which social workers practice is more likely to be guided, albeit unconsciously, by Pauline rather than Kantian principles. Biestek's (1973) conviction that God is the basis of humanity's inalienable worth is contextualised by Cordner's (2002) observation that Christian emphasis upon the moral significance, and love of every individual, was historically unprecedented.

The very notion of a consciousness of selfhood in pre-modern Europe was, Mandler (2004) contends, inordinately circumscribed to a select few, and so continued until well into the nineteenth century, whereas Sorabji (2006) and Thomas (2009) argue for a self-awareness of individuality since antiquity. The modern understanding of the self, Taylor (1989) claims, was expansive, inclusive of *all* humans, and affirmed ordinary life.

The Poor Laws, whilst acknowledging that society had a modicum of responsibility for the alleviation of poverty (de Schweinitz, 1943), were nevertheless underpinned by the assumption that social distress was ultimately explicable as the wages of sin, thereby absolving society from any responsibility or culpability; assistance acted simultaneously as relief and deterrence (Keith-Lucas, 1953). The assailing of pauperism failed to distinguish between moral and economic motives, and resulted in the excision of economics from moral life (Tawney, 1948), and St. Paul's dictum that the poor shall always be with us was taken to heart as a salve to societal conscience. Social work evolved out of religious traditions with an emphasis upon charity (Niebuhr, 1932), understood as 'the love of one's fellow person(s)' (Jones, David, 2004, p. 38), but Jones (1971) asserts that in practice charity was ultimately a means of social control. Moral inclusion was dependent upon economic and social conformity, resulting in a conceptual divide between the deserving and undeserving poor, with penury effectively effacing subjectivity.

The economic upheavals of the 1880s served to undermine the orthodox account of poverty (Booth, 1889), and were a catalyst in the creation of modern-day social work. As Thomas (1983) observes, new attitudes emanate from, and are not external to, the contradictions of existing traditions. That penury came to be understood as a social problem rather than as symptomatic of defects in moral character entailed a marked shift from earlier paternalistic moralism to moral mission (Siporin,

1975). There was a belief in the moral character of the poor, and a moral imagination that saw them as no less deserving of respect (Himmelfarb, 1991). They were seen as individuals within a social matrix, possessing an abundant community life, ingenuity and social strengths (Specht and Courtney, 1995).

In mid-Victorian England, greater emphasis was accorded to the notion of the love of humanity than that previously devoted to the service of God (Woodroofe, 1971), giving impetus to myriad social reforms. Authentic human society was understood to be founded not on economic principles, but upon social relationships (Ruskin, 1900), and the Settlement movement was underpinned by an implicit faith in a common humanity, and a moral commitment to fellowship and equality across the class divide (Terrill, 1974). Among the sundry influences that helped fashion this transformation were Christian notions of agape, caritas and loving-kindness, which served to inspire Christian socialists and reformers (Colloms, 1982), as well as secular socialists (Dennis and Halsey, 1988; Yeo, 1977). This love of humanity was grounded in the particular rather than the abstract – Octavia Hill (Woodroofe, 1971, p. 65) asserted that the principal emphasis was on the creation of 'a solemn sense of relationship', and accentuated the inherent dignity and worth of, and the need to attend to, each and every individual.

Whereas the language of morality is contemporarily supposed to be the exclusive preserve of conservatism, in the Victorian era, which was characterised by a quite remarkable social conscience and compassion, it spanned the political spectrum (Himmelfarb, 1991). Early social workers conveyed to the wider society the essential humanity of the poor as subjective and social beings, and thus members of the moral community; Octavia Hill (Woodroofe, 1971, p. 53) emphasised that each individual 'is a far better judge of it [their life] than we...Our work is rather to bring him to the point of considering and the spirit of judging rightly, rather than to consider and judge for him.' Nevertheless, the emphasis upon individualisation often resulted in individuals being treated judgementally and viewed atomistically (Jones, Gareth, 2004).

At the terminus of the nineteenth-century social work directed its gaze from the amelioration of the individual to the reform of society. Mary Richmond (1922) and Jane Addams (1902), social work's seminal pioneers, embodied social work's dual function, that of a focus on the individual and environmental factors, respectively. In truth neither saw their preferred mode as prescriptively exclusive; Richmond's work with individuals, and her concern for subjectivity (Keith-Lucas, 1953), had a systemic focus, whilst Addams' communities were comprised of flesh and blood individuals. Richmond sees social work as essentially concerned with understanding the human individual within the context of their environment, whilst affirming their uniqueness and the necessity of an individualised response. Both had an implicit respect for the individual, and a commitment to the betterment of society *and* individuals, representing what Pumphrey (1961, p. 64) characterises as the unique characteristics of social work, 'a feeling of obligation *always* to consider social needs when dealing with individuals, and the effect on individuals when dealing with communities'.

The subsequent influence of psychology and psychiatry upon social work from the late 1920s to the 1950s, as opposed to its earlier sociological underpinnings (Alexander, 1972; Keith-Lucas, 1953), more an American than a British phenomenon (Miles, 1954), resulted in an emphasis upon individual inadequacy and psychological adjustment, to the exclusion of structural considerations (Specht and Courtney, 1995), and a disregard of the inherent moral dimension (Siporin, 1982). It nevertheless accorded primacy to subjective experience and accentuated the significance of relationship (Keith-Lucas, 1992; Strean, 1996) and a deeper experience of self (Taft, 1937), and sought to understand motivation rather than explain it away. These aspects were embraced by later humanistic models with their person-centred approaches (Maslow, 1968; Rogers, 1951), the latter having its genesis in both Octavia Hill and Mary Richmond (Rowe, 1996). Reliance upon psychological and sociological theories, to the exclusion of moral philosophy, has resulted in modern-day social work paying mere lip service to respecting individuals whilst adhering, albeit unconsciously, to Marx's injunction to change rather than interpret the world (Wilkes, 1981); the motivation of Graham Greene's (1957, p. 12) Alden Pyle could just as easily be seen as encapsulating that of many contemporary social workers – 'he was determined – I learnt that very soon – to do good, not to any individual person but to a country, a continent, a world. Well, he was in his element now with the whole universe to improve'.

Historically subjectivity has been central to social work's theoretical and practical formulations and its self-understanding, its attention to the individual an essential and fundamental expression of a love of, and service to, humanity. Social work's historical championing of the individual as an absolute end represents a radically compassionate and moral position, conveying to the wider world the common humanity and subjectivity of those with whom it works. In contrast to radical individualism, it acknowledges the inherent relationship between the individual and society and the indispensability of the social matrix, and accordingly values relationship and community. However this notion is often confused and conflated with individualism (Vigilante, 1974), characterised by de Tocqueville (Midgley, 2001, p. 152) as 'a mature and calm feeling, which disposes each member of the community to sever himself from the mass of his fellows ... proceeds from erroneous judgement more than from depraved feelings: it originates as much in deficiencies of mind as in perversity of heart'.

The emphasis accorded to self-determination within social work is often construed as a moral statement about the desirability of personal independence over dependency (Frankel, 1966), with the consequent implication that to be in receipt of care and attention is somehow inherently demeaning, with precedence accorded to those who can be *enabled* (Wilkes, 1985). Margaret Thatcher's remarkable decree that there is no such thing as society, only individuals, represented the apotheosis of individualism, but is belied by the reality that individuals require a social matrix in order to flourish (Taylor, 1985); as the wisdom of the Irish proverb confirms, it is in the shelter of each other that the people live. The notion that personal bonds and mutual interdependence somehow limit our freedom and are inimical to our natures is nonsensical given the social beings we incontrovertibly are – indeed 'For us *bonds* are not just awkward constraints. They are lifelines' (Midgley, 2010, p. 124). It is the nature of our faculties, Midgley (1996a, p. 357) insists, not moral weakness, that underpins our need for personal bonds, for people dissimilar to ourselves, and 'A rational being is someone who sees himself as a unit among others, not as the core of the universe.'

There is, Wilkes (1981) observes, a world of difference between social work understood as a detached and amoral social science, or as related to a metaphysical way of thinking. The former concerns itself with technical know-how and an emphasis upon change and outcomes, whereas the latter is other-centred, engaged in an attentive and respectful exploration of the nature of being. Too little attention is given to reflection upon the *why*, too much accorded to execution of the *how*. Accordingly it will be beneficial to reflect upon how best to understand the individual and their place within the world.

A helpful conceptualisation is provided by Antonaccio (2000, pp. 8–9), who argues that contemporary endeavours to flesh out a plausible and judicious account of moral subjectivity tend to oscillate between liberal and communitarian commitments, to Kantian and Hegelian frameworks respectively; the former 'believe that the self constitutes its own world through its acts and choices apart from determination by the givens of its situation ... the self grounds its identity and its moral claims through some existential decision or transcendental act that escapes the contingencies of the self's situation', whilst the latter 'believe that the aims and purposes of the self are in fact constituted by the givens of its natural, social, and historical existence in particular communities... human subjectivity is not self-constituting but is itself constituted with respect to some antecedent order of value'. The self, Antonaccio observes, is therefore conceptualised as either *unencumbered*, or as *radically situated*.

Liberalism's chief virtue is its insistence that the individual has irreducible value, and communitarianism's that the self is social. Liberalism seeks to contract morality to the private sphere, with goodness a correlate of individual choice, independent of considerations as to what is chosen, whilst communitarianism conceives morality and goodness as inseparable from inherited cultural traditions and the self as having no independent irreducibility (Eagleton, 1997). Because both lack a meaningful sense of a substantial self, the individual is often obscured

> because we are ourselves sunk in a social whole which we allow uncritically to determine our reactions, or because we see each other exclusively as so determined. Or we may fail to see the individual because we are completely enclosed in a fantasy world of our own into which we try to draw things from outside, not grasping their reality and independence.
>
> (Murdoch, 1997, p. 216)

Whereas the world of the ancients conceptualised morality and virtue as independent of will, the modern world deems rationality and will to be constitutive of morality (Taylor, 1989). The upshot of this radical shift, witnessed most clearly in Kant, is that *only* rational beings have unconditional and absolute value. If moral subjectivity is indeed rationality dependent, it needs acknowledging that many humans fail to meet such stringent criteria. Because it is supposed that morality originates in the human will, it cannot in any way be attached to the substance of the world, and accordingly we no longer see ourselves against a background of values and realities that necessarily transcends us; this restricted notion obscures the reality that virtue is connected with unselfishness, objectivity and realism (Murdoch, 1997).

There is an objective reality that necessarily transcends *us*, 'distinct from our imaginings or utterances about it' (Clark, 1997, p. 124), for of necessity 'There must be things single and steady there for us to know, which are separate from the multifarious and shifting world of "becoming". These steady entities are guarantors equally of the unity and objectivity of morals and the reliability of knowledge' (Murdoch, 1977, p. 3). The intimate connection between reality, value and transcendence of self is integral to all great artistic endeavours (Armstrong, 2001; Ruskin, 1995).

Rather than lose much of value in both liberal and communitarian models, it is argued that instead of thinking of morality 'as essentially and by its nature centred on the individual', it is better that we see it 'as part of a general framework of reality which includes the individual' (Murdoch, 1997, p. 68). Reality transcends the individual, but the individual cannot be wholly subsumed in the givens of their situation. And it is the discrete individual, not the rational agent, who is owed respect, who is central to this framework of reality, because consciousness, not will, is the 'fundamental mode or form of moral being' (Murdoch, 1993, p. 171). The conception of an individual, Murdoch (1996, p. 25) claims, is 'inseparable from morality', and this book will argue that such a framework cannot be constricted exclusively to the human world. Whilst observing that it is customary for social workers to view reality as society, Wilkes (1981) insists that given human beings inhabit three worlds, the natural, the social and the world of self, it is preferable that we view reality as the whole of creation. What social work needs is a framework of reality that includes the individual, against a background that acknowledges the natural world and our common animality; an acknowledgement of the moral momentousness of the world outside the self that prevents moral claims from being synonymous with the human will (Antonaccio, 2000).

Philosophy commenced as the quest for, and love of, wisdom, and care of the soul (Clark, 1984), entailing a search for 'self-knowledge as an aspect of understanding the world' (Midgley, 1996c, p. 53). For Aristotle, Plato and Spinoza the aim of knowledge is *contemplation*, 'part of an understanding of life as a whole, out of which a sense of what really mattered in it would become possible. Knowledge indeed had the same goal as love; contemplation was the highest human happiness' (Midgley, 1995b, p. 13). The life of wisdom, Aristotle (1952) insists, constitutes the truly worthwhile life, whilst Plato conceives wisdom as 'the practical intelligence which guides virtuous living' (Annas, 2003, p. 54).

Philosophical dimensions are part of the fabric of social work (Downie and Telfer, 1980; Imre, 1982; Ragg, 1980); social work, Timms (Butrym, 1982, p. 55) declares, is 'primarily neither an applied science nor simple good works but a kind of practical philosophising'.

Social work is fundamentally concerned with knowledge and moral principles, but social work literature has tended to exalt the former whilst perfunctorily doffing its collective cap in the direction of the latter, evidencing of a disposition to dualism in consideration of knowledge and values. Knowledge encompasses far more than mere technique and methodology, and action cannot be an end in itself; we must attend to the *why* before we concern ourselves with the *how*. Knowledge, after all, is not merely a collection of loose facts, but understanding (Midgley, 1990).

Knowledge is neither interchangeable nor synonymous with truth or understanding, for understanding entails the acquirement of knowledge *and* judgement (Stevenson, 1971). Whilst it can furnish us with much of value it cannot of itself instruct us as to *how* we should act or live, or *what* it is that we should do (Millard, 1977). The interposition of judgement between knowledge and understanding reinforces the intrinsic centrality of moral principles, and the immanent relationship between knowledge and morality. It is Stevenson's (1971, p. 232) contention that concentration upon principles and concepts 'enables us to focus and refocus knowledge in a meaningful way'. Knowledge is directed towards contemplation of life as a whole and the acquiring of wisdom, and contemplation entails awe or wonder – 'An essential element in wonder is that we recognize what we see as something that we did not make, cannot fully understand, and acknowledge as containing something greater than ourselves...Knowledge here...is a loving union' (Midgley, 1995b, p. 41). This sense of awe and wonderment cannot however be restricted to our kind:

I believe a leaf of grass is no less than the journey-work of the stars...
And the cow crunching with depress'd head surpasses any statue,
And a mouse is miracle enough to stagger sextillions of infidels.

(Whitman, 1982, p. 217)

An understanding of knowledge as a *loving union* finds its parallel in Murdoch's (1996, p. 30) conviction that 'the central concept of morality is "the individual" thought of as knowable by love', with love being commensurate with knowledge of the individual. Love, Murdoch (1997, pp. 215–16) avers, 'is the perception of individuals...the imaginative recognition of, that is respect for, this otherness'. And it is this love of the individual that transforms self-consciousness into moral consciousness, just as knowledge and morality direct us away from ego, fantasy and illusion, and point us towards reality, goodness and virtue (Murdoch, 1996).

Amazingly, with a few exceptions aside (Hollis, 1967; Morley and Ife, 2002; Wilkes, 1981; Younghusband, 1964), it is a rarity for *love* to be mentioned in contemporary social work literature, a neglect all the odder given its centrality in social work's genesis, and its underpinning of both relationships and community (Lewis et al., 2001). Love respects the independent reality of that which is the object of one's attention and love (Weil, 1952), and is fundamental to social work's philosophy (Siporin, 1975; Tillich, 1962), being a discipline essentially concerned, not with generality, but with a compassionate understanding of, and extension of loving-kindness to, the individual. Social work is intrinsically a process of relationships between selves (Biestek, 1973; Perlman, 1979), and it is therefore advanced that social work is intrinsically concerned with and engaged in acquiring knowledge of the individual, and that the central concept of social work morality is the individual, knowable through love, loving-kindness and loving union.

Levy (1973, p. 34) characterises social work values as providing 'preferred conceptions of people, preferred outcomes for people, and preferred instrumentalities for dealing with people', whilst Banks (1995, p. 4) claims that they refer to 'a set of fundamental moral/ethical principles to which social workers are/should be committed'. The problem that confronts social work is not so much an absence of consensus (Shardlow, 1989) as a dearth of discussion about values, and a lack of conceptual analysis (Timms, 1983). An unwillingness to think, albeit about difficult concepts and dilemmas, and to invite and encourage a dogmatic scepticism about the possibility of a common conceptual value base can hardly be a virtue.

Values provide a background framework and map from which we may gather our bearings and make sense of what it is we are confronted with, and then to bring reflection to bear upon action. The moral life is more aptly conceptualised as piecemeal and continuous, not merely called into action for explicit moral choice (Murdoch, 1996). That is substantially dissimilar from the notion that values are a panacea neatly indexed for all manner of situations, to be referred to as though to a ready reckoner, and applied rote-like to guarantee efficacious outcomes. Complexity is quite distinct from incomprehensibility, calling for a relating of interconnected parts in order that they be understood as a whole.

Precisely because social work engages in the lives of others, social work practice must receive its vindication in moral terms rather than by reference to theory alone (Butrym, 1982; Clark and Asquith, 1985). Outlawing moral conflicts does not serve to rescue us from the horns of dilemma, rather it obfuscates our abilities to effect resolution. Disagreement over emphasis and interpretation of moral principles is quite distinct from denial of the very existence of shared values *per se*:

> Moral disagreement is hardly ever a simple confrontation between opponents who don't share each others' presuppositions at all...Normally a great deal is agreed but not mentioned...Our background of thought is a social network, a vast, complex web of assumptions into which we were born and within which we live. It is not our own private invention or creation.
>
> (Midgley, 1993, p. 151)

Whilst emphases and interpretations have historically shifted, commitment to fundamental values has remained a constant (Siporin, 1975), and Plant et al. (1980) caution that differences about values are more often than not less profound than they initially appear, and disputes about the ends people ought to pursue are quite distinct from what constitutes their needs.

Social workers' predilection for being agents of change (more often than not translating into uncritical adherence to agency function) at the expense of moral reflection merely reinforces the perception that moral considerations are best left to armchair philosophers. The busy practitioner, 'more interested in the state of the roads than in their place on the map' (Tawney, 1930, p. 1), accords scant reflection to the moral principles underlying activity, regarding them as abstractions and distractions from the business of practice. Thus conceptualised, it is exceedingly difficult to envisage how social workers envision that which counts as human good if fact and value are deemed to be fundamentally distinct, for 'knowledge about what goodness means must be at the centre, because it is what shows the point of all other knowledge, indeed of all other activity' (Midgley, 1995b, p. 14).

What is required is conceptual comprehensibility and an intelligent relating of values to practice, for 'To try to use concepts while withdrawing from the conditions of their application, from what conditions their sense, is to saw away the branch on which one is sitting' (Gaita, 1999, p. 270). Theoretical knowledge and practice wisdom are of necessity complementary parts of a whole, rather than competitive paradigms, and both must be informed by underlying moral principles. Practice wisdom should always draw upon contemplation of what is worthy and morally preferable.

It is as nonsensical to suppose that social workers can operate without a value system as it is to spuriously claim that knowledge can be value-free; neutrality truly is a fiction, for *any* social work intervention of necessity involves values (Clark, 2000); indeed *all* our thinking is shaped by our concepts (Midgley, 2001). A separation between knowledge and values has profound epistemological and moral implications, the most disconcerting feature being not the distance between theory and practice, but that between its social values, epistemological assumptions and practice (Clark and Asquith, 1985).

Social work reliance upon knowledge devoted to treatment and outcomes, to the exclusion of an adequate moral philosophy, results in a reductionist understanding of the individual and their experiences, leading to a diminution of respect (Ragg, 1977; Wilkes, 1981). Morality, and moral philosophy, are concerned with what really matters in life, *how* we are to live, and *why*, and the two things required of moral philosophy, Murdoch (1996) contends, are the discoverable attributes of human nature and the commendation of a worthy ethical ideal.

The influence of positivist principles upon social work has served not only to posit a gap between knowledge and values, but to deprecate the standing and validity of the latter, significantly impacting upon the relationship between its articulated value base and practice. Knowledge is conceptualised as that which is scientific *and* value-free, whilst values are characterised as arbitrary preferences, the former *objectively established* whilst the latter are *subjectively preferred* (Gordon, 1965). Whilst there nevertheless remains the elementary expectation that social workers be committed to certain values and ethics, and Imre (1982, pp. 57–8) observes that

> It almost seems as if social work has sought to hold onto values which have been severed from their roots. The reason that these values have not been seriously challenged within the profession is probably because individual social workers have either remained connected to these roots in a personal way and/or have been able to compartmentalize intellectually so that values and the search for scientific knowledge are not seen to be integrally related.

Value-free knowledge is an impossibility, for values are implicit in all theories (Whittaker, 1974), and 'facts are not gathered in a vacuum, but to fill gaps in a world-picture which already exists' (Midgley, 1986, p. 2). The valid point that the proponents of value-free knowledge seek to make concerns a *particular* sense of objectivity, a recognition and exclusion of personal biases or prejudices, and relates to a laudable desire to see things as they *are*, rather than as we believe them to be, or would have them be; 'Truth isn't just *facts*, it's *a mode of being*. It's finding out what's real and responding to it – like when we really see other people and know they exist' (Murdoch, 1987, p. 101). The problematic formulation of objectivity decrees that facts are epistemologically separate from implications of a moral or emotional nature (Clark, 1998a), and seeks to banish subjectivity.

The philosophical rationale for a gap between facts and values was provided by Hume (1958), who contends that whereas reason sheds light upon facts, feelings are the wellspring of values, and morality, in essence, is founded in sentiment, not reason. The attempt to define goodness via the process of deriving values from facts, to argue from *is* to *ought*, was termed a *naturalistic fallacy* by Moore (1903), for, so he argues, good in essence is a non-natural property. Moore supposes that moral judgements are wholly contingent, and is committed to a logical and metaphysical atomism, ignoring the reality that moral judgements cannot be merely random if they are to be comprehended (Midgley, 1980). Whilst it has been pointed out that Moore's conception of goodness entails 'the idea that good was *something* (and so a *kind* of fact)' (Murdoch, 1993, p. 44), the notion that there exists a gap between facts and values has been extremely influential in moral philosophy. In Murdoch's (1997) view, Moore's emphasis upon the human activity of bestowing things with value to the absolute exclusion of any notion of the Good marked *the* definitive breach with metaphysical ethics.

Such disjunction is the causal factor in social work's tendency to assume values as unanalysed givens, allied with a predilection to view values symbolically rather than as the actual fulcrum of daily practice (Vigilante, 1974). The notion that rationality be limited to descriptive, as opposed to evaluative, considerations leads Timms (1983, pp. 135–6) to contend that

> It is possible to use 'value' and 'values' as conversation-terminators...
> 'We must just agree to differ: it's a question of values.'... they simply
> and only report or express people's attitudes for or against some-
> thing... they can only be fought over by a proxy, and never battled
> through to an exchange of ideas and a conversion or change of mind
> or even a conversation.

Such a conceptualisation leads to a moral subjectivism and solipsism, centred upon the atomised individual. Indeed, the whole point of solipsism is the abolition of morality (Murdoch, 1988).

Attempts to define and specify that which is good in human life, and a framework for consequential normative ethics, entail the necessity of a broadening of social work's philosophical base (Imre, 1984). To know what is good and valuable presupposes a relation of values to facts (Nussbaum, 1992), for 'we can indeed only understand our values if we first grasp the given facts about our wants' (Midgley, 1996a, p. 178), and 'if there were no values, there would be no facts' (Clark, 1995b, p. 1). There exists no inherent and insurmountable difficulty in reasoning from facts to values, for it is precisely because our basic structure of wants and needs is *given* that we are capable of discerning what it is that constitutes good for us.

The contemporary assailing of subjective experience had its genesis in the pioneering behavioural psychologist John Watson (1924), who insists that consciousness is illusory and that its disregard guarantees scientific objectivity, whilst Watson's principal behaviourist heir Skinner (1973) conceptualises consciousness and subjectivity (human *and* animal alike) as having incidental epistemological value. This assumption of knowledge *independent* of a knowing subject is belied by the reality that our subjective experiences represent the most substantial aspect of our lives (Birch, 1995; Midgley, 1999). The true paradigm of objectivity is ethical, not epistemological (Eagleton, 2002), objectivity representing 'a decentred openness to the reality of others' (Eagleton, 1997, p. 123), entailing the 'delight in independent realities' (Clark, 1984, p. 114).

The great moral philosophers, Murdoch (1997, p. 64) observes, 'present a total metaphysical picture of which ethics forms a part. The universe, including our own nature, is like *this*, they say'. Metaphysics entails critical reflection upon reality and the construction of images, metaphors and conceptual frameworks (Antonaccio, 2000).

Metaphysics, Imre (1984) argues, is at the very heart of social work, and Wilkes (1981) warns that its absence leads to an emphasis upon change rather than interpretation of our world, an undervaluing of those who remind us of our shared vulnerabilities. The individual, Wilkes (1981, p. 79) insists, ought to be viewed 'not a problem to be solved but a mystery to be apprehended'. When our attention transcends the exclusively human sphere, to matters of animals and the natural world, the necessity of metaphysics is all the more apparent (Dombrowski, 1988; Fox, 1995).

Of late the tide has turned somewhat against moral realism, with the admonition to view all moral beliefs as culturally relative and historically contingent. By way of example, Rorty (Clark, 1998b, pp. 49–50) declares that 'Truth is what it is better for us to believe, rather than the accurate representation of reality', to which Clark (1998b, pp. 49–50, 46) retorts,

> 'we should indeed acknowledge what is true, but that its being true is nothing to do with us.... Those philosophers who equate "truth" merely with what "we" happen to endorse, or doubt the possibility that we are sometimes wrong... have no need to reconsider their own convictions, or imagine that another might be "really right".'

We seek truth, not for truth's sake alone, but *because* it is good (Weil, 1986). Increasingly, Soskice (2004, p. 8) observes, the rationale of the social sciences is beholden to 'successive rhetorics of power and control and are therefore pathways to nihilism'. The alternatives to moral realism will be briefly canvassed in order to exemplify their problematic consequences for both humans and animals.

Moral relativism conceptualises values and morality as unequivocally cultural in origin, and in common with moral subjectivism, eschews universalisable moral principles (Boghossian, 2006; Hughes, 1984). The former posits that we are incapable of relating our moral concepts beyond the frontiers of our own hermetically sealed cultures, and far from transcending morality merely substitutes custom for moral deliberation (Singer, 1993); the latter, as a logical extension of radical individualism, condemns us to mutual moral solitude and incomprehensibility. Moral relativism is not without its redeeming features, it being a clarion call against the assumption that the values of one's culture are unquestionably universalisable. This said, the difficulties in understanding other cultures are unnecessarily compounded by what Midgley (1983b) terms *moral isolationism*, whereby the world is divvied up into disparate, quarantined societies, each with their own unique and unrelatable system of thought. Cultural differences are more a matter of degree than kind, and we are mistaken in overestimating or magnifying dissimilarity (Rachels, 1995). All cultures have a moral concern for what their members make of themselves, and the reason that we can justifiably condemn cultural insensitivity is because of our capacity for making moral judgements in the first instance. Indeed Midgley (1996b, p. 170) contends that to dispense with approval and disapproval, absolutely necessary to our attitudes towards others, would result in an abject failure to treat them *as* people, for 'We need the natural, sincere reactions of those around us if we are to locate ourselves morally or socially at all. They give us our bearings in the world's. Indeed, 'I can be said truly to know who and what I am, only because there are others who can be said truly to know who and what I am' (Macintyre, 1999).

The expression of respect for other cultures is likewise contingent on this capacity (Levy, 2002) given that we cannot respect what is utterly unintelligible, and to treat tolerance as synonymous with respect erodes any substantial conception of what it is to respect others (Cordner, 2002). The particular danger for social work is that it leaves its practitioners open to the charge of being cultural apologists (Meemeduma and Atkinson, 1996). The relativist, Clark (1984, p. 7) observes, 'loses the ability to say that he truly believes anything, for he believes nothing to be true-in-fact. That what he believes he believes to be true-for-him only means that whatever he believes he believes that he believes'. Furthermore, 'if it is true, there is at least one truth that does not depend on us, and so it is false' (Clark, 1986a, p. 56).

Subjectivism reduces morality to personal preference, and so long as one is sincere in one's thoughts or feelings, implies moral infallibility (Regan, 1991). But whereas emphasis upon sincerity entails a focus upon self, an accent upon truth is other centred. Subjectivism sidelines rational deliberation, and we must just agree to disagree – 'when I say that cruelty to animals is wrong I am really only saying that *I* disapprove of cruelty to animals... If this means that I disapprove of cruelty to animals and someone else does not, both statements may be true and so there is nothing to argue about' (Singer, 1984, pp. 6–7).

Relativism and subjectivism not only present significant problems for human beings; postmodernism, which embraces cultural relativism as its foundational concept, all too readily conceptualises the natural world as nothing more than a social construct (Sessions, 1995), ignoring the fact that we construct *conceptions* of things, *not* things (Naess, 1997). Social work is likewise susceptible; for instance, where the reality of child abuse is subordinated to what it means, conceptual analysis obfuscates, or reifies, the lived experience. Social constructionism constricts attention and moral value to humankind, and is committed, Soule (1995) argues, to a relativistic anthropocentrism. Its hubristic claim that reality is what *we* decide entails 'implicitly rejecting any limit on human ambitions. By making human beliefs the final arbiter of reality, they are effectively claiming that nothing exists unless it appears in human consciousness' (Gray, 2002, p. 55).

Postmodernism rejects the moral primacy of consciousness in favour of language (Barthes, 1986; Benhabib, 1992), what Murdoch (1993, p. 193) terms linguistic determinism, whereby 'What is "transcendent" is not the world, but the great sea of language itself which cannot be dominated by the individuals who move or play in it, and who not speak or use language, but are spoken or used by it.' Individuals do not possess an essential identity, personality, subjectivity or a capacity for integration (Parton and Marshall, 1998), subjectivity being characterised by contradiction, precariousness, process and subject to perpetual reconstruction (Weedon, 1987). In its rejection of morality, reality, reason, society and truth (Himmelfarb, 1994; Wolin, 2004), postmodernism detaches us from any pre-existing frame of reference, extolling a fundamental discontinuity of meaning (Foucault, 1970).

Postmodernism can be seen as a reaction to notions of infallible knowledge. It highlights the interrelationship between ideas and the social world (Payne, 1997), and the linkage between knowledge and power, offering a salutary caution against expertism (Trainor, 2002). Social work rightly disavows a *particular* understanding of expertise, for not dissimilar reasons that Shaw (1928) asserts that all professions are conspiracies against the laity. An expert, it's said, is someone who knows more and more about less and less. Nevertheless there is no reason why particular knowledge and skills ought not to be utilised *for* and *with*, rather than *against* or *over* those who seek out the services of a social worker. To lay claim to knowledge and wisdom is clearly distinct from the claim that one knows *all* there is to know about both, or to presume that one is an expert about the lives of others.

There exists a fundamental distinction between what Midgley (1993) terms an *inquiring* and a *dogmatic* scepticism; the former entails due reserve about specific claims until the facts are in and an evaluation can be made, the latter a belief that there exists *no* possibility of answering or resolving moral questions. The dogmatic scepticism that pervades postmodernism presumes that because there is no absolute truth there can be no partial or contingent truths (Himmelfarb, 1994).

Postmodernism has enjoyed appreciable currency within social work (Chambon and Irving, 1994; Fook and Pearce, 1999; Parton and O'Byrne, 2000), but its moral relativism undermines social work practice and core social work ideals, especially social justice (Ife, 1999; Solas, 2002). Its virtual silence on the great liberal ideals of equality, human rights and justice (Eagleton, 1997), and its subversion of subjectivity, ought to alarm social work. In essence, postmodernism is an epistemological rather than moral theory, concerned with 'a theory of knowing, rather than a theory about what sort of society we should have and how people should behave within it' (Fook, 2002, p. 16).

Historically social work unapologetically utilised the language of morality as foundational in justifying its existence, with social workers perceived as moral agents and the moral conscience of society (Siporin, 1982), but contemporarily prefers the terminology of values and ethics (Reid and Popple, 1992). It was not until the twentieth century, Himmelfarb (1995, p. 9) argues, that that the classical virtues were supplanted by the language of values, a consequence of morality being 'thoroughly relativized and subjectified'. That personality has come to supplant character in modern discourse serves to obfuscate one's connections to the wider world (Sennett, 1999), and eclipse the notion of a moral self (Goldstein, 1984).

The oft-declared death of morality evidences a profound misunderstanding of the concept itself:

> the contemporary world seems to favor two supposedly self-evident, if incompatible, propositions on the subject. The first, voiced with either satisfaction or regret, holds that what characterizes Western societies today is the absence of moral life: duty is dead, and in its place we champion something called authenticity. The second proposition tends to be expressed in more imperative tones: it's high time we freed ourselves from the last vestiges of moral oppression, people say. Or: Watch out! Morality is making a comeback!
>
> (Todorov, 2000, p. 285)

Furthermore, Todorov (2000, pp. 286–7) contends that the origin of these propositions rests in misconceptions about both the meaning and gamut of the term moral:

> The first proposition confuses species and genus: from the disappearance of a particular form of morality (the kind rooted in, for lack of a better term, traditional morality), one deduces (falsely) the disappearance of moral life in general ... morality cannot 'disappear' without a radical mutation of the human species.

Even in the death camps, veritable hells on earth, morality survived (Frankl, 1965; Todorov, 2000).

Social work's moral mission is to assist people not only in the resolution of moral dilemmas, but in becoming morally better persons and morally responsible community members (Siporin, 1975); central to this mission is the importance of moral character and virtue in social workers (Clark, 2006), for 'nothing in life is of any value except the attempt to be virtuous' (Murdoch, 1996, p. 87). That social work represents the embodiment of society's conscience is often construed negatively by practitioners, but conscience, Timms (1983) argues, is best conceptualised in terms of content and function, and fidelity to a moral ideal.

One historical source of contemporary unease with the language of morality is the moralising that places absolute responsibility for poverty and social distress upon the purported moral inadequacies of the individual, rather than their being at least in part symptomatic of structural inequalities. Butrym (1982) contends that social workers tend to conflate the notions of moralising and acting morally, and Emmet (1962, p. 169) proposes that in part this results from 'the prevalence of the idea that moral standards are personal, subjective and emotional', and as such are not conducive to rational argument and reflection. In addition, moral judgements tend to be conceptualised as invariably entailing uncharitable or unfavourable judgements on others. In Stalley's (1978, p. 93) view, to be judgemental is, among other things, to engage in authoritarian or dogmatic moralising, as well as the categorisation of individuals that obstructs their being responded to as unique individuals; as a corrective, 'to be non-judgmental is always, so to speak, to "leave morality out of it" '. However those who condemn supposed moralising are seemingly oblivious that they, by their very reproach, are likewise engaged.

Fortunately, morality is not an optional extra, but essential to creatures of our kind:

> Getting right outside morality... would mean losing the basic social network within which we live and communicate with others, including all those others in the past who have formed our culture... [without morality] there is no sense of community with others, no shared wishes, principles, aspirations or ideals, no mutual trust or fellowship with those outside, no preferred set of concepts, nothing agreed on as important.
>
> (Midgley, 1993, p. 10)

Moral judgement entails a continual exercising of rationality, as befits our nature as social and linguistic creatures, and 'is continuous with the rest of life' (Midgley, 1993, p. 26). Midgley (1983b, p. 72) claims that

> When we judge something to be bad or good, better or worse than something else, we are taking it as an example to aim at or avoid. Without opinions of this sort, we would have no framework of comparison for our own policy, no chance of profiting by other people's insights or mistakes. In this vacuum, we could form no judgements on our own actions.

Indeed most of us would not willingly choose to live in, or live with the consequences of, a world in which all human actions were deemed to have moral equivalence.

Our epistemological scepticism invariably results in the dominance of ideology, which Hunt (1978) argues flourishes in social work precisely because philosophy is largely absent from the consideration of substantive issues. There exists, Timms (1983, p. 133) claims, a common assumption in social work circles that *any* observation is so suffused with a person's personal and political values that it makes more sense to inquire about their ideology than their beliefs, 'i.e. the ways in which they seek to confront others about that on which they themselves stand in no need of argument'. Ideology reduces judgement and viewpoints to arbitrary choice or unsubstantiated opinion, and vitiates communication and understanding, and 'once the issues are removed from the sphere of reason, they are liable in practice to be decided by power alone' (Hunt, 1978, p. 20).

As a means of exposing the inadequacy of a gap between fact and value, Stalley (1978) argues that it is well-nigh impossible for social workers to abstain from *any* judgement as to the morality of the actions of those with whom they work, especially in relation to actions that either contribute to or detract from their (or those of other creatures) interests or happiness. A conceptual gap between facts and values leads to a baleful subjectivism and relativism, whereas moral judgements are discoveries about the real world.

Social workers faced with individuals wrestling with moral problems are disrespectful if they act as though none exists. And when attended to, social workers are apt to conceptualise the task at hand as an act of will, as though choice rather than attentiveness were constitutive of our mode of being (Murdoch, 1996). Indeed, were one to proscribe the faculty of moral judgement, then the very notion of moral dilemmas and problems become meaningless, and worse, it 'would be to abandon all attempt to rule our own lives, would amount to a surrender to the judgment of those less humble or cautious than ourselves' (Clark, 1989, p. 174).

What remains to be seen is whether the moral judgement that categorises animals as being outside morality, precluding our knowing them as individuals, is morally justified. Whilst it has been traditional to conceptualise morality as exclusively human in scope, so that human concerns invariably override those of animals, social workers cannot assume that this exclusion is so obvious that it requires no rational justification.

There were, Midgley (2001) maintains, two divergent strands of Enlightenment thinking on the nature of moral considerability, one that that sought to constrict the moral circle, the other that widened its scope, but ever since

All the terms which express that an obligation is serious or binding –
duty, right, law, morality, obligation, justice – have been deliber-
ately narrowed ... to describe only relations holding between free and
rational agents ... It isolates the duties which people owe each other
*merely as thinkers* from those deeper and more general ones which
they owe each other as beings who feel. It cannot, therefore, fail both
to split a man's nature and to isolate him from the rest of the creation
to which he belongs.

(Midgley, 1983c, pp. 166–7, 170)

The traditional argument for the exclusion of animals from the moral
circle correlates moral standing with rationality, witnessed in Kant's
(1964) central moral insight of distinguishing between *subjects* and
*objects*, entailing a normative response that all persons are ends in
themselves. Kant grounds respect for human beings in their rationality
rather than consciousness, as a right rather than consequence, rationality being the faculty that distinguishes and separates us from animals,
contending that 'A rational being must regard himself as ... belonging
to the world of understanding and not to that of the senses', as an
autonomous, not heteronomous, being (Sapontzis, 1987, p. 40). This
is an important distinction, for whilst one may admit consciousness to
animals, it is held that they are irrational beings and therefore beyond
the orbit of direct moral concern. Whilst not dismissive of sentience,
Kant conceptualises it as contributing nothing to the intrinsic worth
of a person. It is from Kantian philosophy that *the* central social work
value, *respect for persons*, is derived, and consequently it is hardly surprising that in social work literature moral value has been conceptualised as
the sole preserve of human beings; for instance, Banks (1995, p. 10)
unequivocally (and characteristically) declares that 'Moral judgements
are about *human welfare.*' To be sure, this is so, but no explanation or
argument is proffered as to why it is, or more to the point, ought to be
exclusively the case. This conceptualisation of morality and subjectivity as referring to normal adult human intelligence not only excludes
animals from the circle of moral considerability but also raises key questions about the status of humans who are philosophically deemed to be
*marginal cases.*

The way in which one conceptualises the subjects of the moral community serves to circumscribe the boundaries of moral considerability
(Johnson, L., 1991; Warren, 1997). Until relatively recently, it appeared
axiomatic that ethics and morality were circumscribedly humanocentric
(MacIver, 1948; Passmore, 1974), although there have been significant

challenges from environmental (Hay, 2002; Nash, 1990) and animal rights (Hursthouse, 2000; Taylor, 1999) philosophies to this human-centred ethic during the latter part of the twentieth century. There has, however, been an uneasy tension between those concerned about animals and the natural world, but to suppose that they are inherently antithetical is patently wrong (Benton, 1993; Callicott, 1992; Jamieson, 2002). Both remind us that we have duties to other beings and entities that extend far beyond the purely rational and human, of a reality beyond ourselves.

Notwithstanding, human welfare and interests have traditionally been held to exhaust our moral deliberations, with the circle of morality conceptualised as being demarcated between ourselves and all other creatures. From such a restrictive and competitive model of moral considerability, Midgley (1983a) argues that it is a short step from *us* and *them*, to the Hobbesian model of *me* and *them*, which invariably misrepresents the inherent complexity of moral claims. The very notion of enlightened self-interest presupposes rationality to be the cornerstone of moral considerability, leaving us without the wherewithal to grapple effectively with the claims of certain human beings, other species and the natural world, as though we were an island of subjects surrounded by a sea of objects.

To have moral standing is to *matter* morally, to some degree, and it imposes limits upon what we consider it right and fitting to do. Moral status endows an individual with a right not to be treated in any which way *we* so desire or decree, to be treated well for *their* sakes (DeGrazia, 2002), bequeathing direct duties. Investigations into membership of the moral community, Regan (1991, p. 21) observes, can be approached by recourse to either *descriptive* or *normative* ethics, the former concerned with '*what is the case* (whether it "ought" to be so or not)', the latter 'with *what ought to be the case* (whether it "is" or not)'.

Whilst observing that definition of the scope of the moral community has historically shifted, Rachels (1995) contends that it refers to beings whose interests are to be taken into account, whilst Regan (1983, p. 152) defines the moral community 'as comprising all those individuals who are of direct moral concern or alternatively, as consisting of all those individuals toward whom moral agents have direct duties'.

There are two questions, Pluhar (1995, p. xiv) asserts, that that are fundamental to moral theory – '(1) What sorts of beings are morally considerable (i.e., proper subjects of our moral concern)? (2) Are all morally considerable beings equally morally significant (i.e., due the same degree of moral respect from us)?'

Rowlands (1998) defines moral considerability as having moral entitlement to equal consideration and respect, whilst Pluhar (1995, pp. 166, xiii) claims that 'To be morally considerable, is to be the sort of being to whom others can have duties', and 'maximum moral respect is due any being, human or nonhuman, who is capable of caring about what befalls him or her'. That said, Goodpaster (1978, pp. 311–12) maintains that it is important to make a distinction between the criteria of *moral considerability* and *moral significance*, the former referring to 'beings who deserve moral consideration in themselves, not simply by reason of their utility to human beings', whilst the latter 'aims at governing *comparative* judgments of moral "weight" in cases of conflict'.

Regan (1983, p. 243) advances two concepts to ground the notion of moral considerability, those being *subjects-of-a-life* and *inherent value*, the former concept defined as follows:

> individuals are subjects-of-a-life if they have beliefs and desires; perception, memory and a sense of the future, including their own future; an emotional life together with feelings of pleasure and pain; preference- and welfare-interests; the ability to initiate action in pursuit of their desires and goals; a psychophysical identity over time; and an individual welfare in the sense that their experiential life fares well or ill for them, logically independent of their utility for others and logically independently of their being the object of anyone else's interests. Those who satisfy the subject-of-a-life criterion themselves have a distinctive kind of value – inherent value – and are not to be viewed or treated as mere receptacles.

Moral agents are morally accountable for their actions and can do wrong, and whilst moral patients can do neither right nor wrong, they have a right to just treatment against moral agents, and the duties we owe them are *direct* because they are individuals who possess inherent value, and as such they are thus owed respect as a matter of justice (Regan, 1983), for 'a morally considerable being is a being to whom moral agents have moral obligations' (Pluhar, 1987, pp. 24–5). The concept of *subject-of-a-life* is applicable to human moral agents *and* human moral patients, as well as, at the least, to all mammals.

The concept of *inherent value*, in Regan's (1983) eyes, has four central components, and is a categorical value – the inherent value of an individual is independent of other's interests, utility, one's character or behaviour, and of a subject's experiences or mental states. Individuals who have inherent value possess moral rights because 'each individual is the subject of a life that is better or worse for that individual' (Regan, 1982, p. 94).

It follows that *all* subjects-of-a-life possess inherent value, but significantly Regan (1983) contends that being a subject-of-a-life is a *sufficient* but not a *necessary* condition for an individual possessing inherent value. In essence, an individual does not have to satisfy all subject-of-a-life criteria to have inherent value. Because moral agents and moral patients can both be harmed in relevantly similar ways, Regan contends that all moral agents and moral patients, whether they are subjects-of-a-life or not, possess inherent value, thereby undermining the traditional distinction.

That the moral community is not as exclusive as traditional philosophy and metaphysics would lead us to believe is evidenced by the reality that we regularly see ourselves as having duties to human beings *independent* of their degree of rationality, a fact that ought to alert us to the possibility that we might have duties elsewhere. If we accord moral consideration to humans who patently fail to meet the traditional, stringent criteria, the continued *a priori* exclusion of all animals on those very same grounds is morally inconsistent. Nor can moral considerability be contingent upon and circumscribed by our capacity for emotional connection (Noddings, 1984), as it provides no rationale to care for those to whom we are indifferent (Regan, 1991).

Murdoch (1996, p. 34) challenges the notion that moral considerability is dependent upon moral agency, asserting that 'the characteristic and proper mark of the active moral agent is a just and loving gaze directed upon an individual reality'. That Murdoch (1996, p. 38) does not believe that this refers implicitly and exclusively to the human individual is borne out by her conviction that the essence of morality is 'attention to individuals, human individuals or individual realities of *other* kinds' [emphasis mine].

Indirect duties represent an attempt to ground duties towards moral patients, whilst explicitly disavowing that we are capable of *directly* wronging or benefiting them, but as Midgley (1983c, p. 171) observes, 'To say "they do not have rights", or "you do not have duties to them" conveys to any ordinary hearer a very simple message; namely, "they do not matter" ... whoever they may be'. There are obvious implications for social work and social work practice; social workers would have to, and be expected to, walk away from the suffering and distress of animals that they encounter, on the grounds that they are of no direct moral significance, that they neither matter, nor are entitled to moral consideration. But social workers cannot, in good conscience, walk away

from the interests of fellow sentient beings, for they *are* morally significant, cannot pretend that their suffering and distress are phantasmal, or
that it is someone else's responsibility. Once concerns and interests are
acknowledged and taken into account, they of necessity must be admitted at both conceptual and practical levels, and accorded intellectual
and moral assent.

Midgley (2001, p. 161) cites a contemporary illustration of this phenomenon, whereby the rationale for opposition to animal abuse rests
upon the improvement of human character:

> the Charity Commissioners in Britain ... warned the Royal Society for
> the Prevention of Cruelty to Animals that it might lose its status as
> a charity if it went on campaigning against forms of animal abuse
> which were advantageous to humans ... campaigns for animal pro-
> tection could only count as charitable in so far as they were aimed at
> 'raising public morality by repressing brutality and cruelty *and thereby
> elevating the human race by stimulating compassion*'.

This attitude is, one suspects, not at all that dissimilar to that of social
work. Social workers might quite legitimately care for and be concerned
about animals in their private lives, but it is considered unbecoming
to allow such attitudes and sentiments to spill over into the public and practice sphere; they ought confine themselves to a concern
for human beings. Any expressions of a modicum of moral consideration for other animals are invariably construed as entailing indirect
duties, and it is precisely this conceptualisation that leaves social workers bereft of guidance in responding to practice dilemmas involving
animals.

In the view of Midgley (1983a), one can discern in the rationalist tradition two dominant tendencies in attitudes towards animals, which she
categorises as *absolute* and *relative dismissal*. The former dismisses expressions of concern as emotional or sentimental, declaring morality to be
a contract between rational beings and we therefore owe them *no* consideration, whilst the latter admits that although animals are entitled to
some consideration, human interests *always* take precedence.

Whereas Hume (1958) excludes animals from the sphere of justice he
nevertheless believes them capable of reasoning (Seidler, 1977), and this
conviction is further developed in its normative implications by both
Bentham and Mill. Counteracting this strand of thought, Kant (1990)
insists that we can have no direct duties towards animals given that
duty itself is a rational bond, whilst Spinoza, although conceding animal

sentience, contends that the distinct nature and emotional structure of humans warrants our unfettered use of animals (Midgley, 1983a).

This traditional constriction of moral considerability to rational and contracting entities, as though it were the solitary mode of ethical relationship, and the rational adult human being the paradigmatic, moral norm is nonsensically narrow, failing as it does to take into account affection, care, love, respect, sympathy, to name but a random quintet. Reason, Salt (1980, p. 114) notes, 'can never be at its best, can never be truly rational, except when it is in perfect harmony with the deep-seated instincts and sympathies which underlie all thought'. Social work clearly does not conceive duty as linked to an individual's degree of competence, and no self-respecting social worker could fail but to object to a state of affairs that conceived babies and other human beings as *marginal* people, or that the boundaries of justice are circumscribed by our rational capacities, for it would represent the antithesis of social work values and practice. It is therefore more than a tad mystifying that social work decrees that that is precisely the manner in which we ought to act in relation to other species.

In the Torah, love for one's fellow specifically includes the stranger (*Deuteronomy* 10:18-19), and the Good Samaritan (*Luke* 10:29-37) does not abandon the stricken stranger he encounters but considers himself to be his brother's keeper, embodying Jesus' injunction and example (*Matthew* 25:35-6). The Good Samaritan epitomises the conviction that one's standing as a moral being is contingent upon the recognition of, and responsibility for, the dependency of others (Bauman, 2001), and highlights the moral imperative of dispensing with the contract notion when we are ascertaining the nature of our duties to our fellows. Traditionally such convictions underpinned social work, as witnessed by Titmuss's (1977) notion of the universal stranger, which argues that we are in a caring relationship, whilst Watson (1980) asserts that social work in essence represents a caring for strangers.

It therefore saddens me to make the observation that the individual attended to by the Good Samaritan was fortunate not to have encountered many a modern-day social worker, for by the time it was ascertained whether or not the 'client' fitted agency guidelines and the proper referral process had been adhered to, the appropriate assessment, contracting and ongoing plan linked to identified goals or outcomes initiated, the poor man would have long given up the ghost.

Ethics, so Schweitzer (1955, p. 239) avers, entails 'Subjective responsibility for all life'. We are not a solipsistic species, and an acknowledgement of this reality requires a radical revision of our moral sensibilities, a recognition that we are not the only species that matter. Unless we are prepared to assert that humankind alone has value in the eyes of God, and nothing in religion demands human uniqueness and animal insignificance as natural corollaries (Linzey and Yamamoto, 1998; Pinches and McDaniel, 1993), or that rationality is the moral attribute par excellence, we ought to acknowledge that our biological continuity infers a shared ethic of compassion for all sentient beings.

The constriction of moral considerability is often predicated on the notion that compassion is a scarce resource, and consequently ought to be exclusively directed towards human beings. The peculiarity of this way of thinking is borne out by the reality that 'compassion does not need to be treated hydraulically in this way, as a rare and irreplaceable fluid, usable only for exceptionally impressive cases. It is a habit or power of the mind, which grows and develops with use' (Midgley, 1983a, p. 31).

Social work's indifference to animals and their well-being and welfare is more explicable, if no less deserving of condemnation, given social work's conceptualisation and understanding of moral considerability. The scenarios presented at the beginning of Chapter 1 could be augmented *ad infinitum* were social workers attentive. Moral indifference is an evil (Baum, 1988; Geras, 1998; Wiesel, 1982), blinding our vision and imaginative capacities and hardening our hearts, and Pinker (1971) suggests that the very existence of social services delineates a compromise between compassion and indifference. An ignorance of morality, Midgley (1993, p. 16) reminds us, 'provides only a reason for *indifference*, for detachment, for not caring about what other people do or suffer'. And such ignorance also ensures that we remain indifferent about the moral status of other animals.

Indifference to animals is underpinned by language, and the moral and psychological device of distancing whereby we steadfastly refuse to think our way into *their* place. It is difficult to overstate the importance that language plays in shaping our attitudes to that which we name or describe; Mola (Adams, 1990, p. 65) argues that our words 'assign status and value'. Our attitudes and conduct towards animals are revealed by the manner in which we speak about them (Dunayer, 2001; Hearne, 1987), and language is regularly and routinely utilised to all but obliterate any residual thoughts of living, sentient beings (Schleifer, 1985), inevitably removing animals from the circle of moral considerability. We disguise their suffering at our hands euphemistically, with animals variously described as 'livestock', 'units of production', 'research tools', 'game', 'meat', and we no longer kill them, rather we 'cull', 'harvest' or 'value add' them. Increasingly in conservation and environmental circles we witness a reification of individual creatures:

> our eye is on the creature itself, but our mind is on the system of inter-actions of which it is the earthly, material embodiment...as long as they are self-renewing, as long as they keep coming forward, we need pay them no heed...An ecological philosophy that tells us to live side by side with other creatures justifies itself by appealing to an idea, an idea of a higher order than any living creature.
>
> (Coetzee, 1999, p. 54)

In failing to respect the otherness and individuality of animals (Plumwood, 1992), 'we are superstitiously Platonic: it is the Idea that is real to us, not the individual suffering entities' (Stephen Clark, 1977, p. 64).

Habitually we employ what Adams (1990) designates as *absent refer-ents*, and language is utilised to mask reality; animals become absent referents in three ways – firstly, and literally, they are absent in the practice of meat eating as a consequence of their deaths; secondly, by definitional means, animals are transformed from living entities into meat; and thirdly, animals are utilised as metaphors for describ-ing human experience. Meat itself is symbolic of our domination of the natural world (Fiddes, 1992; Plumwood, 1997).

Acts of personal and collective cruelty, Haille (1969) contends, are often disguised by means of secrecy and rigid abstraction; the former is often achieved by the utilisation of language specifically employed to disguise reality, whilst the latter refers to the utilisation of terminology that transforms the victim into a creature worlds apart, bereft of either individuality or personality. Both devices are evident in our relations with animals; in the concealment of slaughterhouses (Eisnitz, 1997), the seclusion of vivisection sites from public scrutiny (Ruesch, 1983; Ryder, 1983), and the misinformation and disingenuousness about the actualities of modern factory farming (Foer, 2010; Mason and Singer, 1980). This leads Harrison (1964, p. 144) to make the observation that 'if one person is unkind to an animal it is considered to be cruelty, but where a lot of people are unkind to a lot of animals, especially in the name of commerce, the cruelty is condoned, and, once large sums of money are at stake, will be defended to the last by otherwise intelligent people'. Our moral indifference confirms Bauman's (2000, p. 192) observation that 'morality seems to conform to the law of optical perspective'.

Historically, Thomas (1983) claims that the ethic of human domination served expressly to exclude animals and those considered beast-like from the circle of moral concern. Whilst drawing analogous comparisons between the experience of human beings *in extremis* and the plight of animals (Patterson, 2002; Spiegel, 1989) is almost invariably seen as demeaning, the novelist Isaac Bashevis Singer (1972, p. 257) disagrees:

> As often as Herman had witnessed the slaughter of animals and fish, he always had the same thought; in their behaviour towards creatures, all men were Nazis. The smugness with which men could do to other species as he pleased exemplified the most extreme racist theories, the principle that might is right.

Interned in a Nazi slave labour camp, the philosopher Emmanuel Levinas (1990, p. 153) relates how it was the dog Bobby, 'the last Kantian in Nazi Germany', who reaffirmed the humanity of himself and other prisoners.

Indifference to animals as independent beings is pervasive. In arguing that 'animals lack the individuality which is internal to our sense of human preciousness', Gaita (1991, pp. 120–1) declares that 'An animal's life does not have meaning because an animal cannot live its life deeply or shallowly, lucidly or opaquely, honestly or dishonestly, worthily or unworthily', whilst Tester (1991, p. 42) insists that 'animals *are* blank paper; they are only important because they can tell us something about ourselves... They are nothing other than what we make them.' Gaita's argument takes water when one factors in those human beings who are deemed to be *marginal*; like other animals, their lives nevertheless matter to *them*, and besides, animals can lead good and virtuous lives (Bekoff and Pierce, 2009; Clark, 1985a); Hardy (1976, p. 446), about a bird intentionally blinded, pointedly asks:

> Who hath charity? This bird.
> Who suffereth long and is kind...
> Who hopeth, endureth all things?
> Who thinketh no evil, but sings?
> Who is divine? This bird.

And those of us who share our lives with animals are surely not deluded to have a certain sense of their preciousness.

Tester, in supposing that there is no reality that transcends *us*, seems blissfully unaware as to what this logically commits him to *vis-à-vis* our own species. Casting dogmatic scepticism on our ability to designate cruelty, Tester (1991, p. 207) would have us believe that we are epistemologically inept and moral dilettantes, insisting that 'We cannot know whether the present treatment of animals is preferable to the treatment of two or three centuries ago. There is no supra-historical basis upon which it is viable to ask the question, nor indeed answer it, in anything other than the most trite fashion.' Modern-day tyrants might quite happily avail themselves of Tester's argument, and be inclined likewise to insist that there exists no supra-historical basis upon which we make comparisons about, let alone condemn, the treatment of, say, women, children and other races over time (or be loath to recognise, or admit, the moral validity of such arguments), except in the most banal manner. As Benton (1992, p. 130) comments, 'We can *think* about animals any way we please (so long as we do not mind getting hopelessly muddled), but we cannot *treat* them any way we please: that is a historical *and* biological fact, and also a moral requirement.'

The traditional justification for the exclusion of animals from the moral community, Sapontzis (1987) argues, is dependent upon four principal arguments:

- *reciprocity requirement* (moral rights are in effect the exclusive entitlement of those who possess the capacity to respect the moral rights of others);
- *agency requirement* (entitlement to moral rights is confined to moral agents);
- *relations requirement* (one's entitlement to moral rights against others is dependent upon one's capacity for economic, familial, personal and political relations); and
- *humanist requirement* (moral rights are exclusively concerned with, and confined to, human beings).

In due course, it will be shown that all four requirements signally fail to provide justification for social work's exclusion of animals from its circle of moral considerability.

The *reciprocity requirement* stipulates a symbiotic relationship between rights and duties, but as Sapontzis (1987, p. 144) observes, 'in relations between the powerful and the powerless, it is not reciprocity but moral rights of the powerless against the powerful, without correlative duties of the powerless to the powerful, that are needed for fairness'. Given that human beings are substantially more powerful than animals, Sapontzis maintains that the reciprocity requirement does not preclude the extension of moral rights to animals. Indeed Stephen Clark (1977, p. 21) insists 'those that are weak deserve our especial care', whilst Linzey (1994, p. 28) asserts that 'the weak and the defenceless should not be given equal, but greater consideration. The weak should have moral priority...whenever we find ourselves in a position of power over those who are relatively powerless our moral obligation of generosity increases in proportion.' It is their shared vulnerability, diminished comprehension and articulateness, inability to give or withhold informed consent, moral innocence and their total dependency that secures the special moral status of young children and animals (Linzey, 2009b).

The moral priority due to the weak, the poor and the dispossessed is no newfangled notion (Bauman, 1990; Gutierrez, 1988; *Mark* 5: 1–12), and in status and contract societies it was in the fulfilling of our duties to our social inferiors that we were owed obedience (Clark, 1995a). We would do well to conceptualise moral rights as 'the individual's defence against factitious calculations of the greater good...places in the sun guaranteed to the defenceless' (Stephen Clark, 1977, pp. 22, 29), in order 'To live one's own natural life, to realize one's self...a space in which to live their own lives' (Salt, 1900, pp. 209–10).

The rejection of the reciprocity argument seeks to ground morality in our affections and relatedness, not in our rationality. This in fact is in keeping with the manner in which social workers actually conduct their relationships with fellow human beings, and there exists a strong intellectual and philosophical history that supports this position. Social work does not conceive duties as correlative with reciprocity, for indeed not an insignificant number of those people with whom social workers interact signally fail such a requirement, but are nevertheless entitled to our respect by virtue of their inherent subjectivity. That we extend our moral concern towards the weak and vulnerable within our communities derives from the centrality of natural affections and relationships in our lives (Clark, 1995c).

Our rights, Eagleton (1997, pp. 79–80) surmises, 'may just refer to those human needs and capacities which are so vital for our thriving and well-being'. For social work then to exclude animals on the grounds that they fail to satisfy the reciprocity requirement is to apply moral double standards; it should be obvious that animals besides ourselves also have needs and capacities that are central to their flourishing and good (Grandin, 2009).

Sapontzis (1987) argues that the *agency requirement* self-servingly makes the possession of moral rights dependent upon moral agency, rather than sensation, instinct, emotion or feeling. Whilst acknowledging that human beings are rational in ways that animals are not, and that this makes for an important moral difference, it does not logically follow that an acknowledgement of priority for fully moral agents in non-routine conflict of interest situations thereby commits us to blanket denial of moral rights in moral patients. It is not anthropocentric to state that many animals exhibit behaviours and possess qualities that we would not hesitate to describe as moral or virtuous in our own kind. In primate communities reciprocity and resolution of conflict are of central importance (de Waal, 1996a), and animals routinely display devoted care and love of the young, show loyalty and courage, often exhibit self-denial over food, and the strong often care for the weak. As Sapontzis (1987, p. 28) observes,

> if moral evaluations were confined to the behavioral level of describing what was done, and determining whether it sufficiently resembles paradigm cases of courageous, kind, responsible, and otherwise virtuous action, then it would be clear that animals are capable of moral action. At a behavioral level, a dog's pulling a child to safety is no less a moral action than a human's doing the same thing.

None of this would have surprised Darwin (1936, 1965), for he reminds us that animals possess a rudimentary moral sense, and that human moral agency differs but in degree, and not kind, from that of our animal kin.

Social workers regularly interact with human moral patients, and far from warranting their moral exclusion, their very lack of moral agency entitles them to *greater* consideration, precisely because of their vulnerability; moral agency cannot be held to be morally sovereign over the interests of subjective beings. For this reason, social work cannot resort to the moral agency requirement as a legitimate justification for the non-consideration of the moral interests of our fellow animals.

The *relations requirement* holds that one's entitlement to moral rights is not predicated upon abstract, universal principles, but seeks to draw its sanction from a common morality that is dependent upon one's capacity to enter into economic, familial, personal and political relations with one another. Precisely because animals lack the capacity to enter into relations with human beings, not only are we justified in prioritising human interests, but we have an obligation to do so. In response, Sapontzis (1987, p. 153) contends that this inverts the relationship between morality and community, for 'it is not being part of a community that generates our moral obligations but our sense of moral obligation that generates our community'.

The fact that animals lack this capacity, Sapontzis argues, in no way prevents them from membership in the moral community, and having moral claims against us. In truth, human beings already inhabit what Midgley (1983a) terms the *mixed community*, incorporating myriad domesticated animals, a community of shared lives and interests (Beck and Katcher, 1996; Manning and Serpell, 1994; Podberscek et al., 2000), and this has served to increase our awareness of our biological affinities with them (Serpell, 1986). Indeed Hearne (1987) argues that the relationship between animals and human beings is constituted by a complex, albeit fallible, moral understanding. Society, Clark (1977, p. 35) claims, 'is much more like a household, including different age-groups, ranks and species, and that a similar analogical process reveals the wider Household which is the community of living creatures'. The affection that children feel for the animals with whom they share their lives is not that dissimilar to that felt for their human kith and kin; as Mary Webb's Hazel (1945, p. 79) avows, 'all animals be my brothers and sisters'.

Humans and animals, Benton (1993) argues, are socially and ecologically interdependent; because animals are to some degree constitutive of human societies, any adequate depiction of societies and social relationships must of necessity make reference to animals as embodying social *and* moral relationships. There is a need to distinguish between natural and acquired, as well as negative and positive, obligations; generally speaking, our obligations to non-domestic animals are natural and negative, whilst those appertaining to domesticated animals are acquired and positive in scope (Rowlands, 2002).

Social workers, of all people, ought to be aware of the position of centrality that animals occupy in the lives of many of those with whom they interact in the course of their daily duties, and come to see how odd it would be to stipulate that a being satisfy the relations requirement *before* qualifying for moral consideration of their interests. The households within which they work are necessarily wider in scope than most social workers imagine, precisely because people are not speaking figuratively when they say that animals are family; pets, Sorabji (1993, p. 215) claims, 'are literally *oikeioi* – members of the household'.

Whilst human beings undoubtedly differ in terms of attributes and capacities, social workers nevertheless would be horrified if it were proposed that some humans, on such grounds, were to be denied equal attention and consideration; they would insist, quite rightly, that all human beings have a *right* to be treated respectfully. It is this entitlement to attention and equal consideration of interests that social workers must, for the very same moral reasons, extend to all other animals. For such reasons, social workers cannot utilise the relations requirement to justify the neglect of due moral consideration of the interests of other animals.

Finally, the *humanist requirement*, Sapontzis (1987) argues, draws a hard and fast line around the human species, but if we are good evolutionary humanists, we have no justification for supposing that we are an evolutionary anomaly (de Waal and Tyack, 2003). Ethological studies have challenged not only our preconceptions and prejudices about our fellow animals, but have sought to place us as beings continuous with the natural world. Such findings counterbalance the notion that animals are lawless and brutish, for the roots of morality may be discerned in animals (de Waal, 1996a, 1996b). That we should ever have thought that we alone lived lives of order and virtue seems to have largely stemmed from either ignorance, or our proclivity for hubris. For social work to continue to do so in light of substantial ethological evidence to the contrary, and in post-Darwinian times, appears little more than anthropocentric prejudice and conceit. Absolute dismissal of the moral claims of animals requires justification, not an adherence to *a priori* assumptions, and the onus rests upon those who claim that the moral community and moral considerability are exclusively restricted to human beings. Social work must eventually acknowledge that we ourselves are not moderately similar and somewhat like animals, but that we *are* animals (Midgley, 1996a), and that this reality has normative moral implications.

Given that we *are* animals, utilisation of knowledge from the discipline of ethology, the term utilised to cover all systematic animal behaviour studies, is appropriate, for comparison with fellow species can assist in facilitating understanding of ourselves and our motives. Whilst ethological comparisons make sense only when they are considered 'in the context of the entire character of the species concerned and of the known principles governing resemblances between species', Midgley (1996a, pp. 24, 18) argues that their value lies in placing ourselves within a context of continuity, for '*Understanding* is *relating*...Had we known no other animate life-form than our own, we should have been utterly mysterious to ourselves as a species. And that would have made it immensely harder for us to understand ourselves as individuals too.'

Surprisingly social work has paid little attention to speculations as to the origin of ethics and morality, apart from the theological framework of writers like Biestek (1973). All of us, irrespective of whether or not we see any value in such reflections, operate within frameworks that pre-suppose morality myths; as Murdoch (1997, p. 191) reminds us, 'The mythical is not something "extra"; we live in myth and symbol all the time.' Myths are neither fabrications nor disconnected tales, but 'imag-inative patterns, networks of powerful symbols that suggest particular ways of interpreting the world. They shape its meaning' (Midgley, 2003, p. 1), and that 'When we attend to the range of facts that any partic-ular myth sums up, we are always strongly led to draw the moral that belongs to that myth' (Midgley, 1995a, p. 117). We need to be clear as to what these myths, as to the origins of morality, commit us to holding about human nature, as well as the boundaries of moral considerability.

When considering questions of value, it matters where we begin and what it is that we take to be our background image or picture. Inquiries as to the origin of ethics or morality, Midgley (1991) suggests, confront us with two very different questions; one is concerned with *histori-cal fact*, whilst the other relates to *authority*. Until relatively recently, Midgley (1995a) argues that these questions have been answered in Western culture by reference to two sweeping myths, those being the Christian and social contract myths, both conceiving morality to be an imposition upon human nature, and as constricting moral considerabil-ity to human beings (although it has not been incontrovertible that *all* humans qualify). A third explanation that seeks to include all animals, and that sees morality as inherent in our nature, will be articulated.

In the Christian myth, as the book of *Genesis* reminds us, humankind lived in a paradisiacal state in the Garden of Eden, but the Fall of Humankind, its origin myth, ensured that imperfection entered the nature of human beings. The Christian myth accounts for the ori-gin of morality as a necessary device aimed at realigning our radically imperfect nature with the will of God.

The social contract myth invents a calculating morality to cope with our supposed Fall from the state of idyllic solitude and splendid self-sufficiency, the natural and pre-moral state of humankind, with Rousseau (1950) claiming that it is because we no longer inhabit our naturally solitary and unsocial condition that morality is required to impose peace and order. For Hobbes (1904, pp. 84, 85), our natural state is 'solitary, poore, nasty, brutish and short', one of a 'warre of every man against every man' – Des Pres (1976, p. 142) counters that far from being natural, 'A war of everyone against everyone must be imposed by force' – with morality founded upon enlightened self-interest. Whereas Rousseau argues that is human weakness that forces humankind to be social beings, Hobbes insists that it is fear of others that propels humankind to construct civil society. Whilst Rousseauian egoism maintains that I can identify with and love others, but *only* on the proviso that those others share a likeness to me, Hobbesian egoism proclaims that I have need of others *only* to the extent that they serve my own outward advantage (Midgley, 1996a).

That such people came to place their trust in one another abiding by an agreement in the first instance is perplexing, and the assumption that we are capable of making explicit bargains with fellow rational beings to treat one another decently is inexplicable in the absence of an innate decency (Clark, 1995c). It is folly to suppose the primacy of self-interest to be rational, or that it is irrational to enjoy the company of others or wish them well (Clark, 1999), given that our existence and welfare is ultimately interdependent with that of our fellows.

The egoistic models assume that we were calculating *before* we were social beings (Midgley, 1996a), as though our sociability was a latter adaptation imposed over and above our original solitary and self-interested natures. In truth, we were social long before anything else, for 'we now know – as Hobbes did not – that people are descended from social creatures already provided both with contentiousness and with a strong, subtle, positive sociability to control it' (Midgley, 1983a, p. 86).

Hobbesian enlightened self-interest, Rowlands argues (1998), decrees that contractual commitments apply only to those who represent either a threat or advantage to us, and the unusually weak individuals to whom the strong have no obligations are thereby excluded from any moral consideration. Needless to say, because animals generally pose minimal threat to human beings, and because they cannot contract with us, we have nothing to gain by considering their interests; and as Rowlands comments, the same argument and conclusions pertain to those deemed to be *marginal* humans.

Precisely because animals were traditionally seen as symbols of evil, Midgley (1996a) contends that until relatively recently human beings were afforded a choice between viewing themselves as little or no better than animals (a reductive and pessimistic view), *or* as souls who, whilst embodied, bore no relation to other earthly beings (otherworldly view). In Midgley's (1991, p. 7) opinion,

> On the social contract pattern all animate beings equally were egoists, and human beings were distinctive only in their calculating intelligence... On the religious view, by contrast, the insertion of souls introduced, at a stroke, not just intelligence, but also a vast range of new motivation, most of it altruistic... And today, even among

non-religious thinkers, there is still often found an intense exaltation of human capacities which treats them as something totally different in kind from those of all other animals, to an extent which seems to demand a different, non-terrestrial source.

Our conception of our place in the natural scheme of things has importance for our understanding of ourselves and other creatures. If we consider the natural world to be somehow alien, and to be transcended at all costs, we may well be inclined to believe that only reason can set us free, and that the sooner we shed our animality so much the better. It is Regan's (1991, p. 87) contention that a significant portion of moral and political thought in the Western tradition abstracts human beings from their biological and cultural moorings, with the ubiquitous image being one of an innate selfishness, 'expressed by our standing apart from, rather than in our being a part of, a biological, ecological, or social community'.

Christianity's representation of human beings as uniquely created in the image of God, as souls on a brief terrestrial sojourn, at home neither with our nature nor on earth, or alternatively, Rousseau's portrayal of humankind as essentially solitary and self-contained creatures, and Hobbes' depiction of society as being peopled by atomistic and calculating egoists, might well have retained their appeal and currency were it not for the insights provided by evolutionary theory and ethology. Depicted as 'the creation myth of our age' (Midgley, 1986, p. 30), the theory of evolution has come to occupy a central place in our deliberations, metaphysical and other, about the kind of beings we are. But as Clark (1998b, p. 62) cautions, evolutionary theory, in common with all tales about the past, is never merely descriptive, rather 'it is, implicitly, a political story, slanted to confirm whatever current theorists prefer'.

It can be argued that Darwin's thought has exerted a more persuasive, influential and enduring impact than that of either Marx or Freud. This is especially the case in the search for the origins of morality; whereas Marx sees morality as being in essence nothing more than a bourgeois illusion (Lukes, 1987), Freud conceptualises the individual as being inherently solitary in nature, and culture an imposed and quite alien system (Fromm, 1982; Midgley, 1996b). In contrast, Darwin (1936) proposes that morality arises from *within* our nature, and derives from our innate sociability, countering the notion that we are essentially solitary and independent beings; Butler (1886, p. 394) refers to this piece of metaphysics as 'the speculative absurdity of considering ourselves as single and independent, as having nothing in our nature which has respect to our fellow-creatures'.

Indeed, as Midgley (1994a, pp. 16–17) remarks,

> there had been social mammals who, long before humanity ever existed, had begun to love and help those around them on the ground, not of prudence, but of natural affection. Human beings were descended from these mammals, and quite evidently were not inferior to them in these natural affections...Our primate ancestors...became deeply social before ever they expanded their cerebral cortices.

Given that our human ancestors lived in groups, Singer (1981) contends that it is certain that we would have had to exercise restraint in our behaviour towards our fellows *prior* to us becoming rational beings, and that ethics had its genesis in these pre-human behavioural patterns rather than with the emergence of fully fledged, deliberating, rational human beings. Indeed 'our moral systems are enormously elaborated rationalizations of pre-rational sentiments' (Clark, 1982, p. 109).

Whilst acknowledging that human beings now possess few special instincts, Darwin (1936, p. 481) sees no good reason to suppose that humans did not, from some extremely far-off era, retain some degree of instinctive love and sympathy for their fellows, *in common* with other social animals, who 'are likewise in part impelled by mutual love and sympathy, assisted apparently by some amount of reason'. An instinct and motive to aid one's fellows can be seen as underlying a commitment to the welfare of both individuals and the wider society. For Darwin, morality is our device for resolving conflicts of motive (experienced by all social creatures, both internally and externally), and it this conflict that makes morality necessary. As Midgley (1995a, p. 144) observes, 'Morality, as much as language, seems to be something that could only occur among naturally sociable beings.'

Darwin distinguishes between *strong* and *passing* motives on the one hand, and our *central* and *permanent* social motives on the other; in expanding upon the relationship between social instincts and conscience, Darwin (1936, p. 484) maintains that

> At the moment of action, man will no doubt be apt to follow the stronger impulse; and thought this may occasionally prompt him to the noblest deeds, it will more commonly lead him to gratify his own desires at the expense of other men. But after their gratification when

past and weaker impressions are judged by the ever-enduring social instinct, and by his deep regard for the good opinion of his fellows, retribution will surely come. He will then feel remorse, repentance, regret, or shame; this latter feeling, however, relates almost exclusively to the judgment of others. He will consequently resolve more or less firmly to act differently for the future; and this is conscience; for conscience looks backwards, and serves as a guide for the future.

This capacity for reflection upon our behaviour and motives is innate and inescapable, and is also inherent in social work decision making, where the individual and social aspects are accentuated by considerations of conscience, understood as referring to 'moral feelings *and* to reflection and decision' (Timms, 1983, p. 37). Conscience incorporates both cognitive and emotional components (Garnett, 1969).

But Darwin's account is in no way reductive, for he does not claim for the social affections the mantle of morality; what he does posit is that the social affections, supplemented by increased intelligence (social affections being the matrix, not the consequence, of intelligence), would give rise to reflection upon behaviour (Midgley, 1994a). Darwin (1936, p. 495) maintains that 'the social instincts, – the prime principle of man's constitution – with the aid of active intellectual powers and the effects of habit, naturally lead to the golden rule, "As ye would that men should do to you, do ye to them likewise;" and this lies at the foundation of morality'.

Darwin (1936, pp. 471–2) does not presuppose a radical demarcation between human beings and fellow animals in either the possession of social instincts, or where that sociability may lead, thinking it highly probable 'that any animal whatever, endowed with well-marked social instincts, the parental and filial affections being here included, would inevitably acquire a moral sense or conscience, as soon as its intellectual powers had become as well, or nearly as well developed, as in man'. The revelatory awareness that there are lives besides our own marks the advent of moral consciousness (Clark, 1996a), and it is love of others that transforms self-consciousness into moral consciousness.

The relationship between social instincts, intelligence and morality leads Midgley (1995a, p. 140) to claim that 'The power of thought, if it once makes visible the conflicts of motive that all animals have, must generate morality.' Our social instincts, fortified by both reason and habit, facilitate an extension beyond kin and local concerns, ultimately and logically leading to a widening of the moral circle to include all humans; in contending that the higher moral rules 'are founded on the social instincts, and relate to the welfare of others', Darwin (1936, pp. 491–2) observes that

> Sympathy beyond the confines of man, that is humanity to the lower animals, seems to be one of the latest moral acquisitions . . . This virtue, one of the noblest with which man is endowed, seems to rise incidentally from our sympathies becoming more tender and more widely diffused, until they are extended to all sentient beings.

The theory of evolution, positing as it does a common ancestry, ought to have implications for our view of our animal fellows. It presupposes a radically different metaphysic than we have hitherto been inclined to accept as our guide. In truth, 'we are of a piece with the nature of things' (Clark, 1984, p. 32). The concerns that evolutionary explanations are inherently antithetical to human dignity and serve to degrade human motivation are, as we will discover in the following chapter, not without foundation. But in large part, such disquiet is largely attributable to reductive and misleading impressions of the nature of animals. What Darwin seeks to do is to rectify this insensitive and unenlightened metaphysic, by maintaining that humans differ in *degree* from other animals; as Rachels (1999, p. 174) observes, 'the fundamental reality is best represented by saying that the earth is populated by individuals who resemble each other, and who differ from one another, in myriad ways, rather than saying that the earth is populated by different *kinds* of beings'. Darwinism challenges the traditional notion that human beings and animals inhabit different moral universes, a metaphysic cleaved to with surprising tenacity by many who ostensibly subscribe to evolutionary explanations (Salt, 1921; Shaw, 1949).

Human dignity ought not to be deemed to have such fragility that it invariably demands the derogation of all other species as of no ultimate consequence. What is needed is a recognition that 'Granted that things exist "for their own sake", because God wishes just those things to be, then they are not simply "for us" ' (Clark, 2000a, p. 284). That said, whilst there exists no inherent irreconcilability of evolution and religion (Phipps, 2002; Spencer, 2009), with Darwin (Spencer and Alexander, 2009, p. 13) himself writing in 1879 that 'It seems to me absurd to doubt that a man may be an ardent Theist & an Evolutionist', or that evolutionary explanations of necessity invalidate religious faith (Eagleton, 2009; Haught, 2003), we ought to reject out of hand the notion that evolution sanctions what Midgley (1992, pp. 10, 8) terms *humanolatry*, 'a fantasy that has expanded the notion of "humanism" from a modest, honorable respect for what is good in humanity into a disreputable quasi-religion, exalting us into substitute gods...degrading all other creatures into mere material for our free activity'. As Singer's (1986, p. 33) Joseph Shapiro observes, 'man's treatment of God's creatures makes a mockery of all his ideals and of the whole alleged humanism'.

Humanism has significantly informed social work thought and practice, with its emphasis upon responding to people in a holistic and person-centred manner, with Payne (1997, p. 195) claiming that 'social work values are essentially humanist'. Therein rests both its merit and its flaw, with the scope of morality seen as being exhausted by human concerns; it results in what Clark (2000a, p. 194) terms *universal humanism*, 'a cultural artifact, and not a biological norm...the doctrine that all and only members of our own species merit serious, equal, benevolent concern', whereby 'people have seen themselves as placed, not just at the relative centre of a particular life, but at the absolute, objective centre of everything' (Midgley, 1996c, p. 98).

In line with Darwin's assertion that humankind and other animal species are not radically different in kind, Rachels (1999, p. 174) argues that the way we treat an individual ought to be determined by consideration of that individual's specific characteristics, not by reference to that individual's group membership; this he terms *moral individualism*, 'a view that looks to individual similarities and differences for moral justification'.

Darwin's contributions to our understanding of the origins of morality are significant for ourselves *and* other animals; in observing that the Darwinian picture raises disquieting questions about the extent to which the experiences of animals may approximate our own, Midgley (1989, p. 45) comments that 'The question "Would you like this done to you?" begins to seem increasingly relevant, not on sentimental but on scientific grounds. If we are to dismiss it, we seem now to need much better reasons than those that satisfied some of our ancestors.'

What Darwin shows is how morality could have evolved from, and be perfectly compatible with, our mammalian base, whilst neither undermining our uniqueness (remembering that *all* animals are unique in their own way), nor disparaging our origins:

> Darwin shows how vital our emotional constitution is to all that we most admire, he enables us to accept and celebrate duly this emotional constitution which is so close to that of the other social animals, instead of insisting that everything we value is the work of

that over-strained and hypertrophied cerebral cortex. He leaves us at home with our own nature and on the earth.

(Midgley, 1994a, pp. 17–18)

Perhaps the most important contribution that Darwin has bequeathed to us is that we now know that we are naturally terrestrial beings, and that the natural world is not only our home, but rather the only home we could have, and that all other animals are truly kindred creatures. In providing an increasingly plausible account for the origin of morality, Darwin reminds us that animals ought to be included in the circle of moral considerability.

This chapter has confirmed that social work is an inescapably moral discipline, concerned with subjectivity and notions of what constitutes the good of individuals, and the nature of a good society, and that practitioners' capacity to make moral judgements is an essential corollary of such an enterprise. Ignorance of morality and its origins has been shown to be exceedingly problematic in the human sphere, and antithetical to consideration of the interests of animals.

Knowledge, understood as a prerequisite, but subsidiary nevertheless, to understanding and wisdom, is not the sole preserve of any discipline. *Breadth* is the apposite word (Imre, 1984), for no discipline has a knowledge base that presents as a hermetically sealed circle or island, and no discipline, however specialised or complex it may be, knows all there is to know about their particular field, let alone of the many other fields in our world. Knowledge is an interlocking and cross-fertilising process, akin to overlapping circles (Midgley, 1995b).

The knowledge base of social work should entail the broadest of inquiries, and the following chapter will be exploring disciplines that have either received superficial and perfunctory attention at best, or precipitant dismissal at worst. It is acknowledged that precisely because what is being proposed represents such a radical departure from social work as traditionally conceptualised, a very powerful and detailed case needs to be put forward to do due justice to the issue. If one seeks to change and transform the prevailing worldview of the discipline, one invariably needs to present clear, cogent and compelling arguments, and for this reason I ask for the reader's patience with the inevitable complexity and protracted nature of the subject matter. If we are to change social work's worldview we also need to educate and provide guidance for students and practitioners.

By way of an extended exploration of the nature of human beings and animals (Chapter 3), and the conceptualisation of personhood, moral considerability and respect (Chapter 4), enabling some theoretical and practical implications for social work (Chapter 5), the moral foundation for the inclusion of animals in a social work code of ethics will be grounded (Appendix). The remaining chapters will advance the argument that non-human animals have a moral right to be the subjects of our attention and respect.

# 3
# Social Work's Kindred Creatures: Biological Continuity and Moral Kinship

> We are not just rather like animals, we *are* animals.
>
> *Mary Midgley* (1996a, p. xxxiv)

Nothing is guaranteed to frighten the horses more in the social sciences than suggestions that there is a biological aspect to human nature. Apart from sexuality, which is only too readily identified and attributed to our natural dispositions, without any perceived accompanying diminution of human dignity or freedom, all other human needs are conceptualised as embellishments, created in a higgledy-piggledy manner, specifically without reference to any underlying natural structure.

By way of analogy, we build floor upon floor of our social lives, and endeavour to ascertain understanding without reference to *any* biological base or footings. We are embodied intellects unlike any other being in our world. Humans are social rather than biological beings, shaped by nurture rather than nature; culture is sovereign. But such a view is at odds with the evolutionary continuity, and it will be argued that cultural and biological determinism are equally reductive. As Midgley (1996a, p. 327) observes, 'it is no misfortune to have a specific nature – that freedom, in the sense in which we really value it, does not mean total indeterminacy, still less omnipotence. It means the chance to do *what each of us has it in him to do'*.

Given the crudity of Social Darwinism, and the reductionism of much neo-Darwinism, it is not surprising that social work looks askance at suggestions that biology might inform morality. In observing that evolution represents the contemporary creation myth, Midgley (1986, p. 30) contends that 'By telling us our origins it shapes our views of what we are', but cautions that beyond its official function as biological theory there is a not infrequent assumption that objective scientific knowledge possesses ultimate explanatory powers (Dennett, 1996; Wilson, 1975). Attempts to discern and decipher what is best in human life are philosophical questions and, as Midgley (2003) reminds us, there are many maps of, and windows onto our world and what is needed, above all else, is a scientific pluralism.

The debate, Rachels (1999) argues, tends to oscillate between those who protest that Darwinian theory insidiously weakens traditional morality and for this reason must be rejected, and the many advocates of evolutionary theory who insist that it has no implications and thus poses no threats to traditional morality and values. This impasse serves only to reinforce the perception in the social sciences that that evolutionary theory has negligible epistemological value and no implications for morality *per se*.

This chapter will seek to guide the reader towards a more realistic understanding of the natures of human beings and animals, and the criteria for moral considerability, by critically assessing the notion that human beings are essentially different in kind from all other animals (allied with the scepticism and ambivalence often attached to the notion of human nature), and secondly, the overarching attempts to explain all behaviour (human and non-human) through sociobiological explanations. Both immoderate stances fundamentally misrepresent or distort our understanding of human beings and other animals, for a biological and social account is essential to any comprehensive epistemology and ontology. This done, a middle way that acknowledges that we are biological *and* social creatures will be advanced, one, it will be argued, that is much more respectful of the dignity and value of both the human and non-human animal. Such an exploration will of necessity require careful and thoughtful consideration of key subject matters including the nature of consciousness, instinct, intelligence, language and the relationship between nature and nurture. This integration of the biological and social underpins the moral and intellectual foundations of this book, and it is this integration that provides a new direction for social work and a new code of ethics, allowing the social workers faced with the moral dilemmas in Chapter 1 to know how, what and why to think and act.

Speciesism, 'the widespread discrimination that is practiced by man against the other species' (Ryder, 1983, p. 5), 'has no proper basis in evolutionary biology' (Dawkins, 1989, p. 10), for evolution 'shows us that we ourselves have had ancestors at every level of the animal kingdom, are akin to them all' (Midgley, 1989, p. 45). Biological continuity entails a moral continuum (Ryder, 1983).

Social Darwinism postulated that Darwinian theory provided scientific validation of egoism, and purported to deduce normative ethical and moral implications from the theory of evolution for human society (Hofstadter, 1965; Williams, 1974); 'survival of the fittest', coined by the philosopher Herbert Spencer, was held to be in accordance with the laws of both nature and God (Rachels, 1999), and the theory of natural selection was seen as sanctioning the domination of weak by stronger individuals, classes and species (Ryder, 1989). In truth, it represented a crude travesty of Darwinian theory, and in time fell into disrepute. That said, Darwin's own writings (1936, p. 501) lent some credence to such ideology:

With savages, the weak in body or mind are soon eliminated... We civilised men, on the other hand, do our upmost to check the process of elimination; we build asylums for the imbecile, the maimed and the sick; we institute poor laws; and our medical men exert their utmost skill to save the life of every one to the last moment... Thus the weak members of civilised societies propagate their kind... hardly any one is so ignorant as to allow his worst animals to breed.

Darwin (1936, p. 502) goes on to argue that we nevertheless cannot 'check our [instinct of] sympathy... without deterioration in the noblest part of our nature'.

One can argue that contemporary biological determinism is more sophisticated than its Social Darwinian forbear, and has a less explicit moral tenor. It nevertheless attempts to accord scientific credence to the moral notion that human beings are natural egoists and that the motive of competitiveness underlies all life. Far from having been rendered obsolete, the Social Darwinist metaphysic and psychology continues to provide the backdrop to biological deterministic theories (Midgley, 1995a).

Sociobiology (Dawkins, 1989; Wilson, 1975) and evolutionary psychology (Barkow et al., 1992; Rose and Rose, 2000a) represent two of the most visible and vigorous contemporary incarnations of an atomistic and egoistic doctrine, with Rose (2000) observing that the former mutated into the latter. Sociobiology is defined by Singer (1982, p. 42) 'as the belief that all social behaviour, including that of humans, has a biological basis and is the outcome of an evolutionary process that selects some genes or groups of genes in preference to others'. Nelkin (2000) argues that evolutionary psychology has moved beyond consideration of the relative roles that nature and nurture play in determining human behaviour in order to seek *universal* explanations.

The primary values are held to be survival and reproductive fitness, reducible to the proliferation of genes (characterised by Dawkins (1989, p. 34) as 'the immortals'), and with unadulterated self-interest the omnipotent evolutionary motive, with culture itself transmitted by memes, units of culture that colonise the human mind (Blackmore, 1999; Dennett, 1996). Dawkins (1989, pp. v, 2–3) is adamant that

> We [and all other animals] are survival machines – robot vehicles blindly programmed to preserve the selfish molecules known as genes...a predominant quality to be expected in a selfish gene is ruthless selfishness...Be warned that if you wish, as I do, to build a society in which individuals cooperate generously and unselfishly towards a common good, you can expect little help from biological nature. Let us try to *teach* generosity and altruism, because we are born selfish.

Whilst ostensibly endorsing biological fatalism, decreeing that biology is indeed destiny, biological determinism holds out the promissory hope that we can transcend this destiny via biotechnological elixirs (Rose and Rose, 2000b), with the same fervour once reserved for eugenics (Chesterton, 1922; Hofstadter, 1965). Such a worldview represents the antithesis of social work values, given the centrality of subjectivity and altruism for the moral justification of social work as a discipline.

Human existence is best conceived as being shaped by both nature and nurture, for it is simultaneously biological and social (Rose, 1998). Inordinate emphasis upon genetic explanations results in what Lewontin et al. (1984) term the genetics of blame, with biology continually reinventing itself ideologically as a social weapon (Lewontin, 1977). This results in 'a methodological individualism rooted in biological determinism' that conceptualises society as the regulated battle of all against all, with the individual conceived as 'ontologically prior to the society of which that individual is a part, and that humans have a biologically based nature which society regulates but cannot change' (Rose and Rose, 1982, pp. 7, 10).

Genetic determinism is of especial value in any society seeking to diminish state responsibility for provision of social services, for predisposition, not opportunity, is accorded primacy; existing economic, political social categories are bestowed with a scientific *imprimatur* (Nelkin, 1999). Whilst contemporary sociobiologists deny that their theories have moral or political implications, their writings often imply the opposite (McKinnon, 2006). For example, whilst observing that 'The welfare state is perhaps the greatest altruistic system the animal kingdom has ever known', Dawkins (1989, pp. 117–18) Malthus-like criticises it as an unnatural development – 'Since we humans do not want to return to the old selfish ways where we let the children of too-large families starve to death, we have abolished the family as a unit of economic self-sufficiency, and substituted the state. But the privilege of guaranteed support for children should not be abused' – and takes to task those who encourage the poor – the latter 'probably too ignorant in most cases to be accused of conscious malevolent exploitation' – to breed beyond self-sufficiency. Elsewhere, Ridley (1996) calls for the abolition of welfare payments for sole mothers on the grounds that they are *unnatural*. Social work has a long and honourable history of contesting and countering victim blaming, and ultra-modern variations upon an old theme demand that social work exhibits deftness in redoubling its efforts, in marshalling and deploying alternative defences from within its moral arsenal. We shall now turn our attention to an examination of both the insights and limitations of biological explanations as to our understanding of ourselves (with extended reflection upon human nature) and other animals.

Contemporary evolutionary theory seeks to mould humanity in the Hobbesian image; the gene is personified and bestowed with ultimate meaning, whilst the individual organisms are as so many interchangeable and valueless receptacles (Keller, 1991). In observing that the representation of genes as independent and disconnected intimates an invariably individualistic society, Midgley (1986, p. 46) cautions that 'they are now known to form a most complex system of interdependent parts'. Genes and DNA are not the *only* locus of evolution (Jablonka and Lamb, 2005), and *pace* contemporary evolutionary orthodoxy,

> Far from being isolated in the cell nucleus, magisterially issuing orders by which the rest of the cell is commanded, genes...are in constant dynamic exchange with their cellular environment...So the functioning cell, as a unit, constrains the properties of its individual components. The whole has primacy over its parts.
>
> (Rose, 1997, pp. 125, 169)

The coalescence of egoism and fatalism in sociobiological literature has invariably led the social sciences to regard *all* biological explanations, and an evolutionary context, as uniformly antithetical to human dignity and as having no consequences for our understanding of human beings and their nature (Midgley, 1986). These suspicions are no better exemplified than in sociobiological conceptualisations of altruism, where seemingly altruistic behaviours in fact enhance genetic fitness, that is, the number of offspring an individual may leave behind (Wilson, 1975). The inordinate emphasis accorded to kin as opposed to group selection results, according to Midgley (1996a), in attempts to identify ways in which the individual agent is benefited, and is a problem that arises only for egoists, for it is immediately at odds with the everyday understanding of what constitutes selfishness.

The notion that genetic fitness is the ultimate biological value and altruism merely a derivative function is perplexing, for 'Our normal moralizing does not treat everything as a means to our genetic success' (Clark, 1982, p. 100), and Midgley (1996a) maintains that we would surely consider any person obsessed with maximising descendants either clannish or unhinged. The existence of altruism, and a need to explain it away in other terms, has a long history. Hobbes' decree that we ought invariably to pursue our own interest led him to a psychologically egoistic conception of altruism; his rationale for sympathising with others in their afflictions, and relieving their distress, is that in so doing we alleviate our own anguish (Ryan, 1974). The problem is not that we often feel distress (indeed not to do so would bespeak a dearth of empathic imagination, and a self-centredness), but that this ought to be our exclusive motivation or what we singularly attend to. As social workers we need to attend to others for their sakes, to go beyond placing ourselves in someone else's shoes, and imagine what it would be like to be *them* in *their* shoes.

A significant source of confusion derives from sociobiological technical definitions of terms such as *selfish* and *altruism*. Whilst not officially seen as motives at all, one discerns in sociobiological literature a convergence between selfishness as a state of motivation, and as a genetic phenomenon, a mistaking of metaphors for actual identities, and a simultaneous failure to acknowledge their source (Lewontin, Rose and Kamin, 1984). Clumsily chosen scientific terminology that utilises everyday common meanings has important consequences:

> The fact that 'selfishness' in its ordinary sense is not just the name of a motive but of a fault naturally makes things much worse. To widen the imputation of selfishness is to alter people's view of the human race... Calling someone selfish simply does not mean that they are prudent or successfully self-preserving. It merely says that they are exceptional – and faulty – in having too little care for anybody else.
>
> (Midgley, 1986, p. 117)

The notion that human beings (and animals) are primarily motivated by the proliferation of their genes reiterates that individual beings are ultimately and essentially genetic conduits; for instance, parental-love (Barash, 1980) and mother-love (Daly and Wilson, 1999) are unmasked for being just that, thereby confusing genetic function with genuine psychological motivation by implying that the latter is a deceit. On such a reading, I should not be in the least bit dumbfounded if the day comes when all of my four children spurn my professed and ostensible love and solicitude for them throughout their lives as a grand genetic hoax.

The disparagement of genuine non-reciprocal altruism is a corner-stone of sociobiology; Wilson (1975, p. 157) labels it 'the enemy of civilization', whilst Ghiselin (1974, p. 247) declares 'No hint of gen-uine charity ameliorates our vision of society, once sentimentality has been laid aside. What passes for co-operation turns out to be a mixture of opportunism and exploitation...Scratch an "altruist" and watch a "hypocrite" bleed.'

One discerns in sociobiological literature the salient tendency to suspect the worst of all explicit conscious motivation (Rodd, 1990). The seemingly altruistic behaviour of Mother Teresa is, Wilson (1978) reveals, ultimately explicable as being motivated by her hope for eter-nal life. What appears to be disinterested *caritas* and *agape*, and an embodiment of Jesus' (*John* 15:12–13; *Matthew* 22:39) commandment that we should love others as we love ourselves, is in reality disguised self-interest. On such a reading, solicitude for others is ultimately instru-mental to our ends, and those people with whom social workers work are only there to make *us* feel better. Thus, 'On this view we can do nothing but what satisfies our desires, nothing but use each other for our private profit, even if we sometimes fail to secure the expected gain' (Clark, 1982, p. 57).

Singer (1981, p. 146) argues that philosophers have long been cog-nisant of the tension between self-interested and altruistic motives, and whereas Plato and Kant see a clash between reason and desire, Wilson 'is closer to Hume's view that it is a conflict between self-interested desires and desires like sympathy and benevolence, with reason standing on the sidelines powerless to intervene'. Wilson conflates conscious motiva-tions and biological accounts, assuming that the latter take precedence over the former, and that the former are totally explicable in terms of the latter, but

> It would be absurd to deny that an action is ethical merely because people who carry out the action in fact may in fact gain from it, if

they are not motivated by the prospect of personal gain – and even more absurd if they are not even aware of this prospect...Conscious motivations and biological explanations apply on different levels.

(Singer, 1995, p. 105)

This failure to distinguish between selfishness as a mental state and as a genetic phenomenon is a consequence of the generalised behaviourist distrust of subjectivity (Rodd, 1990), and invariably leads to the viewing of evolutionary function as the supreme motive (Midgley, 1986). The notion that we invariably act so as to further our own interests, be that in the traditional sense, or in the biological sense, and that morality is a ruse of our genes (Ruse and Wilson, 1986), is fallacious. Notwithstanding the ubiquity of the former view, our biology neither compels us to be selfish nor 'precludes the possibility that I would rather that some other creature were content than that I gain some lesser good: I do not seek their good merely as a means to my own happiness – my happiness consists in part in their achieving good' (Clark, 1982, pp. 57–8). Altruistic and cooperative behaviours in both human beings and animals are perfectly compatible with Darwinian principles, and Birch (1999) recounts that Jane Adaams' social work was in part inspired by the Russian anarchist Kropotkin (1990), who conceives mutual aid to be inherent in all of nature and an integral factor in evolution.

In affirming our nature as biological *and* social beings (a position essential for social work, for it cannot think of itself, or those with whom it works, other than in this way), Rodd (1990) asserts that it is more apposite to view human behaviour as subject to sociobiological constraints but mediated by consciousness. This view acts as a corrective to Dawkins' (1989, p. 2) argument that 'Much as we might wish to believe otherwise, universal love and the welfare of species as a whole are concepts which simply do not make evolutionary sense.' The conundrum that surrounds the transmission of altruistic behaviour, which Wilson (1975) describes as *the* principal theoretical dilemma confronting sociobiology, is, according to Midgley (1996a, p. 129), 'solved by showing that it benefits one's kin *and* one's group'; social creatures regularly exhibit tendencies to assist, favour, care for, and take delight in, the young of other parents.

Whilst human beings are predisposed towards kin, reciprocal and group altruism, genuine non-reciprocal altruism is a not an irregular occurrence (Nagel, 1970). Our preference for kin is part of our nature, because 'We are bond-forming creatures, not abstract intellects' (Midgley, 1983a, p. 102), and it is this specitivity which allows for generalisability and the understanding of others (Marris, 1974). This preference 'is the root from which charity grows' (Midgley, 1983a, p. 103), our concern for others having literally familiar origins (Clark, 1997). Rationality does not create the phenomenon of altruism, rather it enables and facilitates its extension to strangers and those currently standing outside our moral circle, plausibly accounting for the ubiquitousness of kindness (George, 2003; Phillips and Taylor, 2009). The predisposition of human beings and other animals to care more for kin than strangers is the consequence of the centrality of personal relationships rather than genetic closeness or calculation (Rodd, 1990; Sharpe, 2005), and '*special relationships* with animals, as with humans, are a basis of positive obligations' (DeGrazia, 1996, p. 274).

Whilst such preferences can often be discriminatory – what Dickens (1986, p. 1240) characterises as 'a perversion of nature in their own contracted sympathies and estimates' – they are neither inherently prejudicial, nor necessarily exclusive (Midgley, 1983a), and Singer (1981, p. 36) argues that 'ethical rules which accept a degree of partiality towards the interests of one's own family may be the best means of promoting the welfare of all families and thus of the entire community'.

While altruistic tendencies must be compatible with gene promotion, these tendencies do not derive from calculation, rather from inherited disposition, and as such are not restricted to kin alone (Midgley, 1984a). Far from reducing individual *fitness*, genes that promote altruistic behaviour are more likely to survive than those promoting selfishness (Singer, 1981).

The criticisms already levelled against the inherent egoism and fatalism with respect to human beings are equally valid with respect to animals. Existence of altruism in animals is not some anthropomorphic fantasy (Bekoff and Pierce, 2009), and if we are prepared to accept the very real possibility of multiple motives rather than a solitary overarching motive, then the behaviours of animals can look markedly different (Masson and McCarthy, 1996).

In exploring the evolutionary origin of affection and love, Eibl-Eibesfeldt (1971), whilst agreeing with Lorenz (1966) that aggression serves an important function in the formation of social bonds, suggests that their roots lie in parental care, in an individualised cherishing of the young, rather than sexuality. The notion that the motives of love and affection are the evolutionary backbone of all social species contradicts the pervasive belief that biological theories justify moral indifference to our treatment of animals (Rodd, 1996). Even a unitary theory such as sociobiology posits a dichotomous metaphysic; human beings retain the vestiges of free will and autonomy, whilst animals are envisaged as genetic automatons, programmed to be thoroughly 'selfish' in the strict biological sense, and thereby incapable of exhibiting self-sacrificing, altruistic behaviour.

Whilst admitting that species bonds are genuine and strong, Midgley (1983a) asserts that they are not invariably exclusive, possessing neither the force nor authority to accord justification for the absolute dismissal of other species. Kin altruism can be utilised to ground human concern for animals (Callicott, 1992), providing us with an awareness of other beings. Rationality enables us to understand the ways in which we are genetically influenced and to challenging that influence (Rodd, 1990), with the ubiquity and universality of the Golden Rule across cultures confirming reason's role in ensuring an impartial ethic (Singer, 1981). This concurs with Darwin's (1936) belief that our social instincts predispose us to have an active concern for and to act, so that others may be benefited, for both human beings and animals possess the capacity to consider the interests of others.

Our ability to extend our natural altruistic dispositions has normative moral implications for our relations with other animals, with Singer (1981, p. 120) contending that 'The only justifiable stopping place for the expansion of altruism is the point at which all whose welfare can be affected by our actions are included within the circle of altruism.' Whilst the case has been made for a moral inclusiveness grounded in our evolutionary continuity, one is nevertheless aware that consideration of biology introduces notions which, on face value at least, appear to weaken if not invalidate what has been argued thus far. It is to these concerns that we now turn our attention.

In countering the belief that biology implicitly endorses competitiveness, Goodwin (1995, p. xii) observes that whereas Darwin assumes that organisms are the basic and irreplaceable units of life, modern biology has substituted genes for organisms; when competition, selfishness and survival are deemed to underpin evolution and complement the values of our culture, 'Both culture and nature then become rooted in similar ways of seeing the world, which are shaped at a deeper level than metaphor by cultural myths, from which the metaphors arise.' This worldview is belied by the reality that

> We are every bit as co-operative as we are competitive; as altruistic as we are selfish; as creative and playful as we are destructive and repetitive. And we are biologically grounded in relationships which operate at all the different levels of our beings as the basis of our

natures as agents of creative evolutionary emergence, a property we share with all other species.

(Goodwin, 1995, p. xiv)

And what's more, 'If organisms are seen as mechanisms, they will be treated as such, and as such we will treat each other' (Goodwin, 1995, p. 215). This is precisely the point that Malik (2000) misconstrues when he fears that acknowledgement of our animal status will invariably lead to human beings being treated as objects, things, or machines. Malik (2000) seeks sanctuary in the Kantian dictum that rationality is the pre-requisite of moral status and subjectivity, and is the *sole* preserve of human beings; whilst animals behave as though they are rational, think-ing and conscious beings, appearances flatter to deceive. They are, in a resurrection of Cartesian metaphysics, merely automatons, and we can therefore discern that Malik's (2000) observations are continuous with the longstanding tradition that morally dismisses non-human animals.

In Malik's eyes evil inheres in treating human beings as we currently see fit to treat animals, losing sight of the fact that to treat *any* con-scious, sentient subject mechanistically is morally pernicious; for 'men exploit animals in much the same way as the rich exploit the proletariat' (Orwell, 1984, p. 459). The attendant dangers in so doing are borne out in Thomas's (1983, pp. 41, 44) study of attitudes towards the natural world in the period from the sixteenth to the nineteenth century:

> this abiding urge to distinguish the human from the animal also had important consequences for relations between men. For, if the essence of humanity was defined as consisting in some specific qual-ity, then it followed that any man who did not display that quality was subhuman, semi-animal... Once perceived as beasts, people were liable to be treated accordingly. The ethic of human domination removed animals from the sphere of human concern. But it also legit-imized the ill-treatment of those humans who were in a supposedly animal condition.

History is replete with such sanctioning of exclusion from the moral community (Benson, 1983; Spiegel, 1989).

The only certain way to guarantee human dignity, Malik (2000, p. 389) insists, is to posit a metaphysical and ontological divide between human beings and all other animals, arguing that 'the less animal we are, the more human we become'. How we can be other than what we *are* Malik does not explain. In reality, the inverse of Malik's prognosis holds true, for to treat non-human subjective beings as automata is inevitably to treat human beings in a like manner (Rosenfield, 1941).

That the elevation of animal-kind is held, seesaw-like, inexorably to detract from humankind is profoundly mystifying, for the reality of social work practice is that when and where children are ill-treated, so are animals (almost invariably), and vice versa (Risley-Curtiss, 2009). If we grant the premise that animals differ but in degree and not in kind, then logic does dictate equality in moral consideration, for compassion and respect are not finite resources. We might just as well suppose that due consideration of, say, women's interests and Third World peoples must diminish the standing of men and those in the First World respectively, or that attention to the poor will unduly impinge upon the well-to-do. Similar arguments were historically advanced to exclude all manner of humans from the circle of moral considerability, arguments against which the social work discipline has historically argued most trenchantly.

*Pace* Tennyson's (1890, p. 261) observation that 'nature is red in tooth and claw' (indeed, if Tennyson's charge be true, it is just as apposite of human behaviour to *all* creatures), mutualism and symbiosis are ubiquitous features of the natural world (Goodwin, 1995); indeed

> competition relevant to natural selection is mainly that within a species than that without...competition, in the vast, impersonal sense required for talking about evolution, goes on, both within species and between them, without the consciousness of those involved in it, and does not at all require what we think of on our tiny scale as deliberate competitive behaviour.
>
> (Midgley, 1983a, p. 24)

Whilst competition, conflict and aggression are an undeniable reality in the animal world, 'they do all this against a wider background of mutual emotional dependence and friendly acceptance...And there surely is every reason to accept that in this matter human beings closely resemble all their nearest relatives' (Midgley, 1995a, p. 134).

By way of summation, Midgley (1984a) argues that the three central failings of sociobiological thought are that it confuses ideas about motive with ones about evolutionary function, it appears to substantially undermine free will, and it fails to make animal comparisons compatible with notions of human dignity and uniqueness. Sociobiology's important contribution, reminding us that we, as animals, are part of the natural world, is diminished by an explicit intent that neurobiology supplant all other disciplines, that ethics be the preserve of biologists, not philosophers (Wilson, 1975, 1978).

Midgley (1995a, p. 73) contends that Wilson is both right *and* wrong; right to draw our attention to the facts about human nature necessary (among many others) for moral choices, but wrong to surmise that such facts can *only* be derived from neurobiology:

> We find out about human nature from a thousand sources, most obviously from everyday life and from history. Without those other sources brain science would not have the concepts and assumptions from which its investigations start. Moreover, it must continually use these outside concepts and assumptions to check the meaning of its work.

Indeed the very idea that sociobiological explanations can categorically encompass and exhaustively account for ethics, or philosophy, is profoundly mistaken (Clark, 2000a; O'Hear, 1997). We should not expect science to definitively capture the complexity of human existence, for science can instruct but not subsume morality (Murdoch, 1996). Similarly, attempts to explain away religious belief as memetic (Dawkins, 1989), as akin to a computer virus (Dawkins, 1992) or a mere function of the brain, is somewhat analogous to expecting the match ball to produce the superlative football of Newcastle United.

There is nothing antithetical to human dignity from biology properly understood, or from an acknowledgement of our ontological continuity. Indeed, as Midgley (2001, p. 185) affirms, 'We know that we belong on this earth. We are not machines or alien beings or disembodied spirits but primates – animals as naturally and incurably dependent on the earthly biosphere as each one of us is dependent on human society.' And it is high time that we recognised that the world does not exist for us alone:

> Let me enjoy the earth no less
> Because the all-enacting Might
> That fashioned forth its loveliness
> Had other aims than my delight.
>
> (Hardy, 1976, p. 238)

The reality of our evolutionary origins is a generally accepted hypothesis, though we are more inclined to the anthropocentric notion that it operates hierarchically, with ourselves at the summit. It's supposed that failure to maintain a metaphysical and ontological divide will result in human beings ceasing to be viewed as subjects and assuredly utilised as objects, but the danger rests not in an acceptance of our ontological continuity but in our treating humans in the manner in which we currently view and treat animals *like* animals, as objects or things; 'the slogan has, precisely, been that humans should not be treated like animals – and by implication animals may be – that is, they may be starved, evicted, imprisoned, tortured, killed whenever it is convenient to "us" ' (Clark, 2000b, pp. 55–6).

Linguistic capacity and competence have ubiquitously been advanced as key distinguishing characteristics between human beings and animals, both ontologically and morally; in Kant's contractual view the meaning of rights and duties is dependent upon a social convention that is expressed linguistically (Midgley, 1996a). This attempt to accord language primacy over and above consciousness is to lose sight of the fact that the latter is constitutive of moral being (Murdoch, 1993). The dogma that *only* humans have minds has its contemporary philosophical advocates, with moral standing deemed contingent upon language, which is in turn dependent upon the possession of self-consciousness (Carruthers, 1992; Frey, 1980; Leahy, 1994). Animals fare little better under sociobiologist Edward Wilson (1975, p. 176), who contends that any attempt to attribute mental criteria to animals 'would be a retreat into mysticism'. In effect, we espouse 'the belief that our particular brand of consciousness makes us uniquely privileged as a species, entitled to evaluate and manage the lives of all others on our own terms' (Mabey, 2005, p. 13).

The traditional view that morality and language are intertwined is 'simply a bad piece of metaphysics... if it were true, there would have to have been a quite advanced point in animal evolution when parents who were merely unconsciousness suddenly had a child which was a fully conscious subject' (Midgley, 1996a, p. 217). It should be obvious that it would be extremely problematic for many humans were moral standing contingent on possession of rationality and language; and it more plausible that we conceptualise the development of consciousness as an outcome of increasing social complexity, and therefore not logically limited or restricted to human beings. Subjectivity exists independently of language possession, and animals' want of human language *elevates* their moral priority (Gandhi, 1984). Thankfully, 'words are not the world... [and] changing the way we speak of things does not change the way things *are*' (Clark, 1986a, p. 53), for 'The universe is wider than our views of it' (Thoreau, 1968, p. 282).

Wittgenstein's (1953, p. 223) oft-quoted aphorism that 'If a lion could talk, we would not understand him' is frequently utilised by contemporary philosophers to deny the attribution of thought, inner life and feeling to all animals *per se* (Clark, 1998c), but this is belied by the fact that we routinely understand the moods and intentions of humans *and* animals via non-verbal communication (Hearne, 1987; Sharpe, 2005), and by Wittgenstein himself (Beardsmore, 1996). Indeed Hearne (1987, pp. 58, 264) contends that to the extent that we deny any capacity for belief, intention and meaning in a dog (for example), we will assuredly not find what we are not looking for, as 'It takes two to conceive', and after all, 'animals are the only non-human Other who answer us'.

Language possession serves a similar function contemporarily as possession of a soul did in bygone ages, securing passage to the circle of moral considerability; Darwin (1936), once again positing difference in degree, not kind, repudiates the notion that language is an essential prerequisite for the possession of rationality. The reality is that any social worker worth their salt would not exclude any human being on the basis of language proficiency, or any other characteristic or attribute, come to think of it; and as Singer (1985, p. 6) observes in relation to supposedly *marginal* human beings, 'The fact that we do not use them as means to our ends indicates that we do not really see decisive moral significance in rationality, or autonomy, or language, or a sense of justice, or any other criteria said to distinguish us from other animals'. Indeed failure to respect animals because they do not share our linguistic capacities is akin to disrespecting the human illiterate (Ingold, 1988a).

In critiquing the notion that language renders all other moral considerations obsolete, Midgley (1996a, p. 225) avers that

> It is not just the fact that a human being talks which gives him a claim to be treated with respect. It is what his talk shows – and he shows the same things in other ways as well, through his actions. If the chimps turn out able to talk well, this can be of enormous interest and tell us much about both their nature and ours. But whatever is true of their moral status is true also, and will continue to be true, of many other fairly advanced creatures. Educated chimps would not form, along with man, an exploitative elite, exalted above all other life-forms as subject above object. The only intelligible arrangement is to regard

*all* animals as subjects of some kind, though with a life that varies greatly in its kind and degree of complexity.

In effect this asks no more of us than that which we already do in relation to *all* human beings, as an article of social work faith (in theory at least), be extended to all other animals; that is, quintessentially, moral consideration of the interests of *all* subjective creatures.

What immediately strikes one in reading Darwin is his readiness to attribute emotions, intelligence – to even the humble earthworm (Darwin, 1945) – and motives to animals. His convictions about language would surely have received further substantiation and validation by the latter-day attempts to teach human language to chimpanzees, gorillas and orangutans, wherein comparisons are made with language development in human children (Cavalieri and Singer, 1993; Linden, 1976).

From the nineteenth century onwards it was assumed that the correct and knowledgeable way to view animals was objectively and scientifically (Rollin, 1990; Walker, 1983), and

> In place of a natural world redolent with human analogy and symbolic meaning, and sensitive to man's behaviour...a detached natural scene to be viewed and studied by the observer from the outside, as if by peering through a window, in the secure knowledge that the objects of contemplation inhabited a separate realm, offering no omens or signs, without human meaning or significance.
>
> (Thomas, 1983, p. 89)

The self-satisfied superiority of supposedly civilised societies was sharply contrasted with the reverence exhibited in the animistic and anthropomorphic attitudes of putatively primitive peoples (Midgley, 1994b).

A fairly standard riposte to the claim that animals are indeed conscious beings is the accusation of anthropomorphism (Kennedy, 1992), relating more often than not to a blanket denial of consciousness or sentience rather than identification of the attribution of erroneous feelings and thoughts (Midgley, 1983a), and in its extreme form outlawing the ascription of *any* human attribute to animals (Crocker, 1984; Fisher, 1990).

The *a priori* decree that only human beings are conscious, or that animals can only have mental experiences on the proviso that they be the selfsame as our own, leads Griffin (1981, p. 124) to declare that 'It is this conceit which is truly anthropomorphic.' We form our opinions of other species by comparison to our aptitudes, interpreting difference as confirmation of inferiority, and underestimating complexity (Coy, 1988), and we are mistaken to view animals as failed attempts at humanity, for 'Present-day species are in no way more 'primitive' than men…[they] are our contemporaries' (Stephen Clark, 1977, p. 111). More often than not respect is due to *unlikeness*; were it otherwise it would often be difficult to see what similarities exist between social workers and many of the people with whom they work, and social work education seeks to cultivate and effect an engagement with difference.

It is incomprehensible to suppose that experience could have emanated out of that which is singularly physical, for consciousness itself is an evolved endowment (Griffin, 1981). A more plausible hypothesis is that consciousness across species arose as a consequence of the demands and increasing complexity of social life, and has a biological function for carrying out introspective psychology (Humphrey, 1976, 1979). The capacity for doing psychology, Humphrey (1979, p. 58) argues, is a characteristic of all social animals, and the need for introspection is precisely why consciousness is required, for 'different kinds of knowledge entail different ways of knowing'.

A necessary interdependent relationship between intelligence and our sociability seems sensible, but the belief that it is the exclusive province of human beings is evolutionarily nonsensical. Nevertheless the tendency to accord pride of place to intelligence is pervasive, with Wilson (1975, p. 381) hypothesising that intelligence is *the* 'impelling force' that generated our sociability. However, to suppose that intelligence made its grand entry at some stage in the evolution of human beings *as* human beings is surely to put the cart before the horse. A far more satisfying explanation, given our evolutionary continuity, is Midgley's (1996a, p. 130) observation that

> There might, perhaps, have been an intelligent species somewhere which did not develop direct social impulses at all, but depended for all its social activity on calculation of consequences. We are not it…Insofar as there is one 'impelling force', it is sociability. From that comes increasing power of communication, which provides the matrix for intelligence.

The very sentiments that are intrinsic to human beings are not derivative of our intellectual prowess; rationality itself 'includes a definite structure of preferences, a priority system based on feeling…[and] is not peculiar to the human race, but is also found in the higher animals' (Midgley, 1996a, p. 256).

What of Humphrey's (1979) assertion that different kinds of knowledge of necessity entail different ways of knowing? By way of observation, Nagel (1974, p. 440) observes that our predilection for explanation of incomprehensibility in familiar and accepted terminology results in improbable representations of mentality:

> the analogical form of the English expression 'what is it *like*' is misleading. It does not mean 'what (in our experience) it *resembles*,' but rather 'how is it for the subject himself'... The subjective character of the experience of a person deaf and blind from birth is not accessible to me, for example, nor presumably is mine to him. This does not prevent us each from believing that the other's experience has such a subjective character.

Nor ought it to diminish one iota the moral importance of subjectivity, and a preparedness not merely to place oneself in another's shoes, to assume that an experiential subject dwells therein. That this is unremarkable is because 'we know from the inside what it is like to be an *animal*' (Clark, 2003, p. 198).

The desire for explanation in well-known and well-understood terminology is of particular hindrance in connection with the inquiry into animals' minds (Bavidge and Ground, 1994; Radner and Radner, 1996); to see animals solely in our terms is to obscure efforts to know them as they *are*, to encounter and respect their otherness, their unlikeness:

> I am looking out of my window in an anxious and resentful state of mind, oblivious of my surroundings, brooding perhaps on some damage done to my prestige. Then suddenly I observe a hovering kestrel. In a moment everything is altered. The brooding self with its hurt vanity has disappeared. There is nothing now but the kestrel. And when I return to thinking of the other matter it seems less important...we take a self-forgetful pleasure in the sheer alien pointless independent existence of animals, birds, stones and trees. 'Not how the world is, but that it is, is the mystical'.
>
> (Murdoch, 1996, pp. 84–5)

'Pointless' in the sense that its existence has value *independent* of human calculations:

Glory be to God for dappled things –
For skies of couple-colour as a brinded cow;
For rose-moles all in stipple upon trout that swim;
Fresh-firecoal chestnut-falls; finches' wings.

(Hopkins, 1967, p. 63)

These are reflective of what Clark (1986b, p. 1049) terms 'intelligent piety', whereby 'truth is known through love, awe, worship'.

The tendency that leads us to invariably understand animals by means of reference to human similitude is critiqued by Bavidge and Ground (1994, p. 170), who maintain that we must distinguish between the statements:

a) Animal pain is like ours.

b) Like us, animals experience pain.

The first is analogical...we are saying that we must always be engaged in the analogical modelling of the subjective qualities of their experience on the subjective qualities of our own. All our thought and talk about animal minds is an exercise in *comparative objective phenomenology*.

Reliance upon analogy alone hinders openness to understanding the subjective and expressive lives of animals, inhibiting the ascription of consciousness and psychological states. As to where else we may commence, if not from our own experience, Bavidge and Ground (1994, pp. 170–1) suggest that 'If there is a comparative element in our thinking it involves large-scale comparisons between our lives and the lives of other species...'Like us, animals experience pain', that is, our lives share common shapes'.

Rather than insisting that all consciousness must be identical with that to be found in animals of our kind, consciousness is better conceptualised as having diverse manifestations (Rogers, 1997). Whilst analogy can indeed be misleading and obscure an animal's viewpoint, at a fundamental level it is indispensable; Dawkins (1985, p. 40) argues that analogy which makes recourse to biological knowledge of animals, as opposed to seeing them as alike ourselves with add-ons of feather and fur, 'is the only kind of analogy which, in the end, will give us any real hope of being able to unlock other species from their skins and of beginning to see the world through not just our eyes but theirs as well'.

To speak of motives and animals in the same breath is routinely deemed an open-and-shut case of anthropomorphism, or in sociobiology's view, delusory (Rodd, 1990). In examining the central importance of motives for ourselves and animals, Midgley (1996a, pp. 105–6) alleges that this accusation is thoroughly misplaced, for motives 'are not the names of hypothetical inner states, but of major patterns in anyone's life, the signs of which are regular and visible ... Both with animals and with men, we respond to the feelings and intentions we read in an action, not only to the action itself'.

Difference is difference, rather than automatically being a reason for the attribution of inferiority, and indifference. Noting that reason is the faculty of the human mind deemed paramount, Darwin (1936, pp. 453, 494–5) makes the observation that

> the more the habits of any particular animal are studied by a naturalist, the more he attributes to reason and less to unlearnt instincts ... the difference in mind between man and the higher animals, great as it is, certainly is one of degree and not kind. We have seen that the senses and intuitions, the various emotions and faculties, such as love, memory, attention, curiosity, imitation, reason, &c., of which man boasts, may be found in an incipient, or even sometimes in a well-developed condition, in the lower animals.

Although 'We are oblivious to other forms of awareness and sentience around us' (Peterson and Goodall, 1993, p. 180), we are not islands of consciousness in an otherwise unconscious world; it is far more economical to explain the complex behaviours we observe in animals in mentalistic terms, for we inhabit a shared world that 'contains a cornucopia of minds' (Bavidge and Ground, 1994, p. 2). Insight into the minds of others occurs via an innate imaginative sympathy (Harris, 1991), and 'As social creatures ourselves, we perceive and respond to consciousness in others in a special way' (Midgley, 1983a, p. 91).

The methodological ruling that consciousness is the sole preserve of human beings has been extremely influential, with animal subjectivity viewed as 'an extravagant metaphysical hypothesis' (Midgley, 1983a, p. 134). Whilst folk psychology readily attributes mental states to animals (Rips and Conrad, 1989), it was but a short step from an objective methodology to a methodological ruling when it came to the study of animals and the natural world (Merchant, 1990). Given the difficulty of rigorous substantiation, Griffin (1992) remarks that scientists repeatedly fall back on the time-honoured precept that *all* animal behaviour, ingenuity notwithstanding, is unconscious. Conversely, the existence of mental states in animals accords with biological and psychological continuity, with the difference in consciousness between humans and other animals being a matter of degree, not kind. Given the dominant religious and secular traditions in the Western world, Midgley (1996a) remarks that the prevalence of a dismissive attitude towards animal intelligence is unsurprising, for wherever there exists a tradition that a particular subject matters not one jot, it is not only exceedingly difficult to eradicate, but acts to all intents and purposes as a conversation terminator.

For the greater part of the twentieth century, science and behaviourist psychology sought to simplify existence by adopting either an atheistic or agnostic orientation towards consciousness in humans and animals (Boakes, 1984; Rollin, 1990), both thoroughly *un*Darwinian positions. In deeming it 'impossible to think objectively about subjectivity', Midgley (2001, p. 8) asserts that a naïve dogmatic materialism decreed that consciousness was not only trivial, but to all intents and purposes had no place in our world, and it was this assumption which led behaviourist psychologists to their outlandish and ultimately unworkable deductions:

> The first step is to measure whatever can be easily measured. This is OK as far as it goes. The second step is to disregard that which can't be measured or to give it an arbitrary quantitative value. This is artificial and misleading. The third step is to presume that what can't be measured isn't very important. This is blindness. The fourth step is to say that what can't be easily measured really doesn't exist. This is suicide.
>
> (Yankelovitch, in Griffin, 1981, p. 112)

And whilst contemporarily seen as antithetical to a realistic representation of human existence, it is assumed to be an apposite explanation for our understanding of animal existence (Hearne, 1987), 'not only as a methodological device to delay too ready an assumption of intuitive insight, but as an ontological dogma' (Clark, 1982, pp. 14–15). Attempts to apply such a methodology to the social work sphere, for instance in the areas of child protection and mental illness (to name but two examples), would result in absurdity, and be ethically reprehensible.

Whilst the issue of consciousness is an area of inquiry that has always intrigued and enthralled philosophers, Bavidge and Ground (1994, p. 8) observe that (a sprinkling of dissenters aside) 'philosophers interested in the nature of mind have done philosophy almost as if there weren't any animals', with the inordinate focus upon the division between the conscious and non-conscious marginalising interest in its diversity. What has occurred is a prevalent tendency to view consciousness as a substance, and '*This is what stops it being accepted as a normal aspect of mental activity, an emergent capacity acquired naturally by social creatures during the regular course of evolution*' (Midgley, 2001, p. 110).

The ever-circumspect Darwin continually sought to counter and rebut the notion of evolutionary discontinuity. By way of addressing the mental powers of animals, specifically in relation to abstraction, general conceptions, self-consciousness and mental individuality, Darwin (1936, p. 446) holds that 'there is no fundamental difference between man and the higher mammals in their mental faculties'.

Can it be that language *causes* consciousness? Budiansky (1999, pp. 18, 193) thinks so, pronouncing that language '*is* a discontinuity', being not one of degree or quantity but quality, 'for whether or not language causes consciousness, language is so intimately tied to consciousness that the two seem inseparable'. Language faculties, Chomsky (1968) contends, are innate and uniquely human, and evolutionarily distinct from animal communication. Language joins consciousness as an attribute that we are encouraged to believe appeared miraculously and uniquely in the human species, examples of what Chomsky (1968, p. 62) terms 'true emergence'. Chomsky adopts the innatist view that the brains of humans and animals differ because of specific genetic differences between species, but Goodwin (1988) insists that this ignores the genetic similitude between humans and chimpanzees. Evolution, Midgley (1983a, p. 140) asserts, *demands* continuity:

Tradition, following Descartes, has tied them (uses of consciousness) closely to our higher intellectual capacities... before any of these faculties can be used, those possessing them must already have all the very complex emotional and perceptual adaptations which make it possible to live harmoniously together. But no creature could even start on the arduous road that leads to this condition unless it was already conscious.

Chomsky places a chasm between language and gestures, thereby ignoring their inherent linkage, 'for speech makes sense only for a species that is already constantly communicating by gestures' (Midgley, 1996a, p. 243). The focus upon linguistic discourse invariably obscures the richness of communicative gestures, many of which we share in common with animals; 'Neither with dog nor human do we need words to reveal to us what expressive and interpretative capacities far older and far deeper than words make clear immediately' (Midgley, 1983a, p. 59). It is our shared evolutionary background, not language, that facilitates mutual understanding.

The pervasive belief that intelligence requires linguistic competency is mistaken; intelligence is more plausibly linked to a being's aptitude for appropriate and flexible responsiveness to specific situational problems, and an ability to so conceptualise the situation (Dupre, 1990). And if our capacity for language is dependent upon innate faculties, as seems the only sensible explanation, it is nonsensical to suppose 'that speech could have originated among creatures which had no understanding, no concepts, no emotions, no beliefs and no desires' (Midgley, 1983a, p. 60). The notion that language pre-dates consciousness, that thought was an impossibility prior to speech is, as Clark (1997, p. 145) illustrates, 'to make the acquisition of speech a standing miracle in every growing child, and in the first beginnings of the human kind... how do individuals begin to speak without ever having thought before they spoke?'

Rather than relying upon language possession as *the* differentiating moral attribute, Carruthers (1992, p. 171) distinguishes between consciousness and reflexive consciousness, arguing that whilst animals are conscious in the sense that they are aware of the world that surrounds them and their own bodies, they do not have conscious mental states – 'If consciousness is like the turning on of a light, then it may be that their lives are nothing but darkness.' They are seen as 'unfortunate demicreatures inhabiting a flickering world of dulled consciousness and automated desire' (Peterson and Goodall, 1993, p. 180). Whilst our dogs Cilla, Lucy, Simone and Tessa, and Jayke the cat, for instance, may be aware of Clarabelle, our Jersey cow, waiting for her morning molasses at our paddock fence, they are incapable of reflecting upon that awareness. The upshot is that 'Once animals have been ontologically darkened in this way, the conclusion that they lack meaningful moral status becomes tempting' (Johnson, E., 1991, p. 192).

Although conscious and non-conscious mental states, distinguished as reflective awareness and perceptual consciousness (McDaniel, 1989), more commonly understood as self-aware and aware (Sharpe, 2005), differ considerably, the difference is not absolute. A stipulation that one be able to engage in rumination upon one's thoughts as a prerequisite for a capacity to feel, to be aware or to possess moral considerability, would leave a significant number of humans adrift; subjective beings have value in and of themselves, irrespective of whether or not those beings possess the capacity to reflect upon that value, for 'interests need not be consciously felt in order to be one's own' (McDaniel, 1989, p. 62).

Conscious experience, Chalmers (1995, p. 80) suggests, is simultaneously 'the most familiar thing in the world and the most mysterious'. However, consciousness is somewhat demystified if we understand it, not as though we ourselves were freaks of nature, but 'as simply an intensification of life – a stronger form of that power to use and respond to one's surroundings which is characteristic of all living things' (Midgley, 2002, p. 11), or as 'the subjectivity of being' (Clark, 1984, p. 29).

It is ironic, Ingold (1988b, p. 96) remarks, that that we demand of animals that they be conscious and aware in *all* their activities, and proceed to act from rationally deliberated plans, when humans seldom do so in the course of everyday life – 'Animals act as conscious, intentional agents, much as we do; that is, their actions are directed by practical consciousness. The difference is simply that we are able to isolate separate intentions from the stream of consciousness, to focus attention on them, and to articulate them in discourse.' Indeed, when it comes to our capacity for learning, it remains a distinct advantage that we do not think about a significant amount of the things that we do, for it thereby allows us to direct our concentration elsewhere (Medawar, 1957). That introspection is contemporarily isolated and elevated as the litmus test for both humanity *and* moral status intrigues Griffin (1984), given that since the behaviourist era it has been accorded negligible attention by psychologists in the human sphere. The very absence of a compelling case for the denial of consciousness in animals makes the consideration of their well-being and suffering morally obligatory.

Carruthers (1992) avers that the mental states of *all* animals are non-conscious, and they are therefore not only excluded from any moral considerability in their own right, but are precluded from even indirect moral consideration, whilst Budiansky (1999, p. 193) pronounces that their purported inability to be conscious of consciousness entails that 'sentience is not sentience, and pain isn't even pain', as though even we have to *think* about our pain for it to take on heightened moral significance; as Fodor (1999, p. 12) remarks, 'Surely what matters to whether it's all right for me to step on the cat's tail is primarily whether it hurts him, not what he thinks about it.'

Carruthers (1992, p. 195) further claims that

> we find it intuitively abhorrent that the lives or suffering of animals should be weighed against the lives or sufferings of human beings...the beliefs in question are so deeply embedded in our moral thinking that it might be more reasonable to do without any theory of morality at all, than to accept one that would accord animals equal moral standing with ourselves.

Indeed Carruthers insists that to feel sympathy for animals is proof positive of misplaced priorities and the result of our own rational imperfection. We can likewise console ourselves that other humans are less troubled by their circumstances than we delicate souls would be were we in their shoes. Carruthers' absolute dismissal is conspicuously at odds both with Darwin's explication of the origins of morality, as well as morality being something that is naturally expansive in scope.

Though expressive of admirable states of character, Carruthers (1992, p. 168) admonishes against concern for animal welfare, urging that 'Our response to animal lovers should not be "If it upsets you, don't think about it", but rather "If it upsets you, think about something more important".' Carruthers is endorsing the moral injunctive equivalent of laying back and thinking of humans, and his whole conceptualisation of compassion as a scarce resource that we clasp close to our hearts for dear, needless-to-say, *human*, life, completely misrepresents the nature of compassion.

Indeed, the annals of social work provide ready to hand confirmation that compassion has never been conceptualised thus, and this fact ought to counter the notion that social work cannot be interested in the welfare and well-being of animals so long as there are human beings deserving of our attention; given that from antiquity we have been told that the poor will always be with us, the implication is somewhat obvious. The latter argument is surely belied by the reality that social work has consistently adapted and expanded in accordance with circumstance and need, especially since the midpoint of the twentieth century.

Even if we were to concede that all animals are not self-conscious, that does not banish them from the sphere of moral considerability, any more than it does those humans who lack reflexive consciousness. It needs noting that it's equally difficult to prove that animals are not conscious beings as it is to prove the affirmative (Griffin, 1984).

It may well be that animal consciousness is invariably located in the here and now, but even *if* this is the case, it is unparsimonious to conclude that what seems to our everyday senses to be best described as consciousness ought more aptly be described as the actions of automatons. Were it otherwise, it would be a matter of utter indifference what state any animal experienced, which is patently nonsensical to anyone who has had a modicum of dealings with animals. Why this is so is borne out by Midgley's (1983a, p. 92) assertion that 'A conscious being is one which can *mind* what happens to it, which *prefers* some things to others, which can be pleased or pained, can suffer or enjoy', and Rollin's (1990, p. 72) certitude that pain represents 'the most obviously morally relevant mode of consciousness'.

In nonlinguistic and unreflective animals, it is perfectly intelligible to employ the terminology of preference (Jeffrey, 1985). The ability of other species to interact with human beings strongly correlates, in the view of Coy (1988), with a sense of self, as the ability to distinguish self from non-self is to be found in varying degrees in individuals of many species, as well as a capacity for being able to impute actions to their conspecifics by conscious projection. Indeed Sharpe (2005, p. 101) argues that 'consciousness of self can also be defined as awareness of oneself as one among other selves'. Once we acknowledge as much and emerge from the Cartesian metaphysical straitjacket, we enter a qualitatively and quantitatively different moral landscape. We are not the only beings that matter, and when confronted with animal suffering we do not need to lay back and console ourselves with supposedly worthier thoughts. Generally speaking, social workers are neither unaware of, nor unfamiliar with, animal suffering, abuse and neglect, as witnessed by the social workers in the practice dilemmas in chapter 1; the unfortunate fact however is that, generally speaking, they just do not know what or how to think, or what to do about it.

Our confidence in attributing conscious mental states to animals, Searle (1992, p. 73) maintains, 'isn't just because the dog behaves in a certain way that is appropriate to having conscious mental states, but also because I can see that the causal basis of the behaviour in the dog's physiology is relevantly like my own'. What Searle sees, social workers routinely see – for instance, in exhibited behaviours that unmistakably evidence suffering in abused and neglected young children. Once we acknowledge this fact, it seems mystifying that it should ever have been otherwise.

The biological continuity of mental capacities lends credence to the notion that subjectivity is to be found in other animals, as does similarity between animals and ourselves in nervous systems, biochemistry and pain-experiencing behaviour (Rollin, 1990); 'If consciousness exists in human beings, it must have evolved; and if human consciousness evolved, it must have had some precedent' (Radner and Radner, 1996, p. 119). Nothing logically compels us, Dawkins (1998, pp. 176–7) observes, to make the leap from acknowledging behavioural similarities to affirming that animals therefore share consciousness with human beings, however

> Our near-certainty about shared [human] experiences is based, amongst other things, on a mixture of the complexity of their behaviour, their ability to 'think' intelligently and on their being able to demonstrate to us that they have a point of view in which what happens to them *matters* to them. We now know that these three attributes – complexity, thinking and minding about the world – are also present in other species. The conclusion that they, too, are consciously aware is therefore compelling.

The unmistakable conclusion and normative implication is that social work cannot walk away or adopt an attitude of moral indifference to the well-being of any conscious creature, for whilst consciousness may well differ in degree, it does not differ in kind. Acknowledgement of consciousness in other species is not merely descriptive, but has normative implications:

> It seems to follow that there is a conceptual link between admitting consciousness and accepting some social duties. These duties find their general *raison d'etre* here, and not at the point where we first detect reason, self-consciousness or speech. Of course these more impressive qualities impose important further duties of their own, but they are not the frontier of all social duty.
>
> (Midgley, 1983a, p. 92)

Budiansky's (1999) position is essentially a reworking of Frey's (1980, p. 163) acknowledgement that animals experience pain, 'or as I would prefer to say, since I think they lack this concept, have unpleasant sensations'. To place Frey's observation in context, he bases his denial of the attribution of rights to animals on the claim that animals cannot have interests; they are incapable of having interests because they cannot have desires or emotions; they cannot have desires or emotions because they cannot have the thoughts required for them; and as the final and telling piece in the jigsaw, the reason they cannot have such thoughts is because they do not possess a language.

The denial of emotions and desires to animals is of particular moral consequence, for as Sprigge (1979, p. 94) maintains, 'a genuine grasp of the fact that another feels an emotion necessarily involves taking account of it, in deciding what to do, in the same kind of way as one would if one directly felt it oneself', and as a consequence when one assumes one's own interests have a significance that the interests of others do not, one is acting in ignorance, since 'one is refusing to recognize that the feelings of others are realities in just the same way that one's own are'. Historically, sentience has always been seen to have primary moral significance, although membership in the sentient order has always been contested. Unless we cede the proviso that the infliction of suffering always requires exceedingly good justification (Lewis, 1947), undertaken specifically to advance the best interests of the individual, we invariably fail to have a full appreciation of the reasons as to why it is immoral to treat human beings in certain ways.

The reason why a being's capacity to experience suffering, rather than pleasure, has occupied a more central place in moral thought is, in Midgley's (1983a) opinion, due to the fact that the duties it entails have *greater* urgency. Indeed it is Darwin's conviction that a creature's sentience, rather than our own sympathy towards it, that ought to determine moral considerability. By way of explicating the centrality of sentience in our moral deliberations, Dawkins (1985, pp. 27–8) observes that the acknowledgement of subjective experiences in other humans acts as necessary and fundamental moral inhibition against the infliction of pain and suffering on others:

> This is one of the cornerstones of our ideas about what is right and what is wrong...Then we come to the boundary of our own species. No longer do we have words. No longer do we have the high degree of similarity of anatomy, physiology and behaviour. But that is no reason to assume that they are any more locked inside their skins than are members of our species.

Indeed, subjective assumptions of similarity and a capacity for empathy are the cornerstones of human relationships; 'Emotionally, it calls for an intense, unstoppable interest in what others are thinking and feeling' (Midgley, 1995a, p. 143). Each of us as human beings only has *direct* knowledge of our own private consciousness, but this does not commit nor condemn us to solipsism; we ought to be grateful that we are not mutually invisible (Laing, 1967). As Clark (1984, p. 16) avows, solipsism cannot even be articulated as a metaphysical proposition, for

To say anything, even to say that I am the only reality, is to say that something is true, and would be true even if I did not know it...once it is admitted (as I must) that there is a world distinct from my experience of, and my thought about, that world, I can admit the existence of other perspectives also, other minds.

We routinely attribute subjective experiences to other humans despite any absolute certainty, so we ought not to be dogmatic naysayers in relation to the experiences of fellow animals (Dawkins, 1985). The capacity for empathy is essential for the practice of social work and an understanding of others, and social workers are cognisant of the fact that those with whom they work inhabit a shared, not alien, world; it facilitates a capacity to engage with others, as well as the ability to imagine what their experiences might be like for them. Our capacity to sympathise with our fellow creatures 'is indeed a cognitive function, part of the way we know the world' (Clark, 1997, p. 48).

*Pace* Budiansky, Primatt (1992, p. 21) maintains that 'Pain is pain...and the creature that suffers it, whether man or beast, being sensible of the misery of it whilst it lasts, suffers *evil*.' As an aside, in nineteenth century medicine sentience was often denied in those deemed lesser humans (Pernick, 1985). Sentience is morally central, the '*prerequisite* for having interests at all' (Singer, 1976, p. 9), and 'the bedrock of our morality' (Ryder, 1989, p. 325). If pain is *not* an evil, then arguments both against and for vivisection fail (Lewis, 1947).

Fundamentally, social work responds initially to suffering, not happiness; one's capacity to suffer makes one a legitimate subject of social work attention and concern. What this book asks of social workers is in this sense not a radical departure from the discipline's central duty, rather that they ought to respond to suffering irrespective of species membership. The moral point of consciousness rests in the fact that 'to be conscious is to be a subject, not just an object' (Midgley, 1995c, p. 72), and to evict and banish animal consciousness from the real world 'abandons the profound respect for life and for the experience of others which has always lain at the root of morality' (Midgley, 1995a, p. 31).

The notion that one must have the capacity to reflect upon one's thoughts about one's experiences in order to feel in the first instance is contrary to our everyday intuitions and modes of thinking. We do not attribute either consciousness or sentience to the appliances and machinery that surround our everyday lives, but were a local vet to

operate upon our dog without anaesthetic we would quite rightly interpret the howls and tremors as evidence of suffering. Nor do we, when a
baby falls headfirst onto the floor, deliberate as to whether or not that
individual is experiencing a sensation or is actually in pain, or whether
they possess the capacity to bring reflective consciousness to bear on
their experience; its absence in an abused or neglected baby, infant
or toddler, or in a profoundly mentally disturbed or senile individual
serves to *heighten* concern. The absence of consciousness of the consciousness of pain is all the more reason to accord beings *greater* rather
than lesser significance, for 'If they are in pain, their whole universe is
pain; there is no horizon; they *are* their pain' (Rollin, 1990, p. 144),
and the capacity to feign pain, shared by many species besides our
own, presupposes the capacity to experience the real thing (Bateson,
1973).

Furthermore, most people would consider anyone who classified the
presenting behaviours as sensations as opposed to pain, or that pain is
not pain, as plain odd or heartless.

Nor ought we to give any credence to the curious notion that
sentience is contingent upon language possession, for animals readily
provide vocal and bodily evidence for their being in pain – substantial
evidence points to its existence in fish as well (Braithwaite, 2010):

> I see that animals are in distress, and the notional addition of a lan-
> guage would not assist my perception. 'I am in pain', after all, only
> 'replaces' a cry of anguish, which must be recognized as such before
> words may be taught or learned. 'I am in pain' does not explicate,
> identify or prove the existence of that anguish ... It may be that ani-
> mals feel less distress, or fewer distresses, than we do, though we
> might remember that the same has been said of the human poor,
> and of the racially distinct.
>
> (Stephen Clark, 1977, p. 39)

We don't demand uniform consciousness (or rationality or language
proficiency for that matter) as a requirement for its existence within
our own species, but this penchant for a definitive distinguishing characteristic assumes that what is specified as constituting good must of
necessity be exclusive to a species. We lose sight of the fact that *every*
species is unique in its own way:

> Whatever the bird is, is perfect in the bird.
>
> (Wright, 2004, p. 1)

And furthermore:

> They know Earth-secrets that know not I.
>
> (Hardy, 1976, p. 147)

Given that essence cannot be expressed in a single differentia, but rather in complexity, Midgley (1996a, p. 321) avers that the inquiry should be one based upon what distinguishes us *among*, not from, other animals; speech, rationality and culture, the traditional distinguishing characteristics of human beings, do not somehow stand in opposition to our nature, but rather are 'continuous with and growing out of it'. Our capacities are dependent upon our animality (MacIntyre, 1999).

For understandable reasons social work, and the social sciences generally, have shied away from notions that human dignity rests within our natural dispositions and underlying structure of needs. The concept of human nature has almost invariably been construed as committing one to deterministic and fatalistic implications that negate or truncate free will and autonomy. More often than not freedom is conceptualised as open-ended, but the 'equation of freedom with unlimited possibility, with random action, is very ill-founded' (Clark, 1982, p. 38). In arguing the case for the compatibility of our evolutionary origins and human freedom, Midgley (1995a, p. 164) observes that '*Unlike machines, which typically have a single, fixed function, evolved organisms have a plurality of aims, held together flexibly in a complex but versatile system. It is only this second, complex arrangement that could make our kind of freedom possible at all.*'

The concept of human nature is ubiquitous in both everyday life and theoretical discourse, and it would appear that most people have at the very least a rudimentary notion, perhaps more intuitive than considered, perhaps more speculative than indubitable, of what it is that defines the human condition. How often do we hear it said that an almost infinite variety of human behaviours, good or otherwise, can be assigned to, or made explicable in terms of, human nature. To do without any conception of human nature is akin to discarding all maps that might otherwise serve to guide our steps, however falteringly and tentatively, towards self-understanding and the understanding of others. As Geldard (1995, p. 7) comments, 'Every true philosopher must eventually lay out a vision of human nature, one that answers our crucial questions: Who are we? What is the meaning of our existence? How are we to live?'

It will be helpful to ascertain or establish what it is we are making reference to when we speak of the nature of a species being – defined as 'the particular combination of qualities belonging to a person or thing by birth or constitution; native or inherent character...the instincts or inherent tendencies directing conduct' (Macquarie Dictionary, 1990, p. 1140), Midgley (1996a, p. 58) claims that it 'consists in a certain range of powers and tendencies, a repertoire, inherited and forming a fairly firm characteristic pattern, though conditions after birth may vary the details quite a lot'. Both definitions make the uncontroversial assumption that the nature of a being refers to certain qualities, characteristics and tendencies that are shared in common by all members of the species in question, albeit in differing degrees, and enables us to have insight into understanding exactly what is conducive to the well-being and good of, or to the detriment of, that species being.

In the human species, for instance, there are certain qualities and tendencies, an underlying structure of needs and dispositions, that enable us to categorically state that their frustration constitutes an evil; for instance, because human beings are social and naturally bond-forming creatures, it is an evil to subject any human to perpetual solitary confinement, or to deny parents and children ongoing contact and interaction. We are not indeterminate beings, nor are we able to recreate the human condition at will. This reality merely sets the parameters within which we as a species operate, the range of possibilities open to us, and does not imperil free will or human dignity, or inevitably engender fatalistic determinism. The greatest threat to all things we cherish about ourselves comes not from the concept of human nature, but from the fanciful and nonsensical notion that we are infinitely malleable creatures; constructivist accounts of the family, for instance, seek to deny the reality and centrality of our biological connections (Almond, 2006).

Geras (1998, p. 87) suggests that there are four distinctive views of human nature, the first holding that human nature is intrinsically evil, and

> 'need not require that people are by nature wholly, or even that they are all inordinately, evil...more likely...that impulses toward evil are sufficiently strong and extensive in humankind that they can never be lastingly pacified, and must continue to produce horrors of one sort and another on both a small and large scale'.

The second view holds that human nature is intrinsically good. The historical roots of this view are grounded in the Enlightenment belief in the infinite perfectibility of human beings, a belief subsequently endorsed and embraced by socialism, and Geras (1998, p. 87) observes that human capacity for evil 'is to be seen as less typical of, or less powerful within, the species, as adventitious and removable, as due possibly to the corrupting influence of bad circumstances or inadequate education; where the potentiality for good is more integral, more deeply laid'.

The third view holds that human nature is intrinsically blank; Geras (1998, p. 86) conceptualises this position as one that assumes that 'the social conditions, or relations or institutions, fully determine the traits borne by any given group of social agents'.

The fourth view claims that human nature is intrinsically mixed, whereby the potentialities for good and evil in human beings are both 'a permanent part of the constitution of humankind...and neither kind of potentiality is held to bulk so large as to be overwhelming or to render the other, whether now or at any time, inconsequential or null' (Geras, 1998, pp. 84, 88).

The latter view of human nature is more in keeping with the realities of social work practice experience, which tells us that human beings are extremely complex creatures, endowed with mixed motivation for good and evil. Nothing in this counteracts nor negates human capacity for change and growth, or the exercising of free will. Indeed that is a more apposite charge to level against the three preceding views; the fatalism of the first view would, if true, serve to vitiate much social work, and in all likelihood engender alternating attitudes of contempt and pity for those with whom social workers work, whilst the initial promissory optimism kindled by claims of infinite perfectibility or infinite malleability, that human evil has merely external causes, patently fails to adequately account for human evil, let alone extend due justice to the complexity of human nature and individual moral responsibility.

The view that human nature is intrinsically mixed is predicated upon the assumption that human tendencies towards good and evil are 'permanent features of our natures, realities to be negotiated, lived with, if possible understood – and if possible tilted toward the more benign and admirable, and tilted that way as far as possible' (Geras, 1998, p. 89). This view concurs with the notion that human nature, far from committing us to illusory visions of absolute autonomy or fatalism, is the raw material with which we must work and over which we must exercise rational and moral discernment. As Midgley (1995a, p. 20) reminds us, 'Inner conflict itself, of a kind to be expected in an evolved creature, is thus to be seen as central to freedom and to the morality by which we try to manage it'; therein resides our freedom. We are, after all, biological *and* social animals, and there are distinct natural limits to what we can make of ourselves, notwithstanding Foucault's (1983, p. 216) contention that our aim in life is 'not to discover what we are, but to refuse what we are'.

To see ourselves as infinitely malleable creatures, or as the existentialists would have it, that existence precedes essence (Sartre, 1957), there being no human nature (Sartre, 1958), is to conceive our dignity and freedom as dependent upon the transcendence of our terrestriality. The notion that natural needs are the antithesis of freedom, and that 'our ideals should in no way be limited by our needs' (Battersby, 1980, p. 273), or that facts about human nature are but 'prevailing tendencies' and Man ' "is" what he decides to do' (Cottingham, 1983, pp. 466, 469), are both ubiquitous and misleading. It is all the more perplexing considering that Cottingham (1983, p. 465) acknowledges that humans are animals 'with a specific genetic inheritance', and yet disavows that there exists *any* universal facts about human nature that enable us to identify that which constitutes natural goods. Our social and natural inheritance is the realm within which we live (Midgley, 1996a), and what is needed, Antonaccio (2000, p. 8) insists, is 'a conception of the self that both acknowledges our freedom *and* allows that freedom to be conditioned by the contingent facts of our situation'.

The penchant for viewing human nature dualistically has continued largely unabated from Platonic and Christian traditions right up to and including the present day; Midgley (1995a, pp. 129–30) recalls that 'The idea of animality as a foreign principle inside us, alien to all admirable human qualities, is an old one, often used to dramatize psychological conflicts as raging between the soul and "the beast within".' The latter, essentially an attempt to solve the problem of evil (Midgley, 1996a), is almost invariably linked with vices and wickedness, our constructed and accumulative mythology of animals serving not only to obscure and misrepresent the natures of animals, but to deem any comparison or similitude as demeaning; as Midgley (Wynne-Tyson, 1985, p. 203) observes, 'Someone who has buttressed his sense of his own dignity by allowing no dignity at all to anybody else, naturally feels any suggestion of a relationship with those others as intolerably degrading.'

Dissenting, Whitman (1982, p. 218) muses:

> I think I could turn and live with animals, they are so placid and self-contain'd/
> I stand and look at them long and long.

In noting that 'Misdeeds and moral illusions keep close company', Benson (1983, pp. 79–80) declares that 'Cut adrift from the demands of discovering and responding to animals as they present themselves to us, we are free to invent their natures, floating at the impulse of need and fantasy from one false image to another.' We come to see animals not as they are, but as objects upon which we project our own anxieties, desires and prejudices. The belief that the devil often manifested in animal form (Sheehan, 1991), or in monsters, who were themselves seen as the appalling products of the mixing of species, served to make the human/animal boundary problematic (Davidson, 1991), and reinforced the notion that any blurring was to be interpreted as a threat to human dignity and uniqueness. Notions of the beast within serve to mask human propensity for irrationality and moral corruption (Benson, 1983); human capacity for evil leads Dostoevsky (1952) to remark that if the Devil be a human myth, then we have assuredly created that being in *our* very own image and likeness. The assumption that there is a lawless beast within human beings when none exists outside of us results in making 'man anxious to exaggerate his differences from all other species and to ground all activities he values in capacities unshared by animals', whereas 'It would be more natural to say that the beast within gives us partial order; the task of conceptual thought will only be to complete it' (Midgley, 1996a, pp. 197, 40). What anthropology did for the myth of the Savage, Midgley (1973) notes, ethology does with the Beast myth, irrefutably authenticating the reality of social bonds among social animals, and that this sociability is not merely a means to an end (Midgley, 1995a).

Because, or so we are told, animals are beings without will (or soul, mind, consciousness, language, *ad infinitum*) and slaves of instinct, human beings are as different from animals as chalk from cheese. In truth, they have natures of their own, with underlying structures of needs and natural dispositions, many of which we share in common. In noting that ethological literature, spanning many decades, identifies patterns of resemblance and continuity between human beings and animals, Midgley (1983a, p. 14) argues that

> The more we know about their detailed behaviour, the clearer and more interesting this continuity becomes. Accordingly, to grasp more fully how their lives work inevitably gives us a sense of fellowship with them. And at this point an emotional and practical concern does naturally tend to join the speculative one.

And whilst ethological literature is seen as predominantly descriptive in essence, Bekoff (2002) calls for a deep reflective ethology that will cultivate greater awareness as to the nature of our moral obligations towards other animals. The notion that there are genetic causes of human behaviour is distinct from the implausibly sweeping and fatalistic claim that the behaviour of human beings is uniformly explicable in genetic terms. What is being suggested is a far more modest assertion in terms of Darwinian theory; if we understand instinct as 'like collective habits of the species...shaped by experience through many generations' (Sheldrake, 2000, p. 13), as incorporating a large collection of abilities (Hearne, 1987), as referring to 'a disposition' (Midgley, 1996a, p. 51), and not a unitary concept, for 'Rather than being a modular Swiss army knife, it is better likened to a kitchen drawer containing a heterogenous collection of implements with different uses' (Bateson, 2000, p. 170), then we ought not to be unduly alarmed.

The notion of instinct is often seen as the fundamental distinguishing characteristic between humans and non-human animals, leading to the metaphysic that humans and animals necessarily inhabit disparate ontological and moral universes. Whereas human beings are *infinitely* free and disembodied intellects, animals are *slavishly* captive to their instincts. Whereas Descartes holds to a theory of innate knowledge, Locke believes that we are all born without knowledge but with instincts, which predispose us to act and feel in certain ways; Watson, the founding father of behaviourism, took Locke's theory to its extreme conclusion by positing that we are all born *without* instincts (Midgley, 1996a). *Pace* Watson, Blake (Ackroyd, 1999, p. 11) believes that 'Man Brings All that he has or Can have Into the World with him. Man is Born like a Garden ready Planted & Sown.'

Such an ontological schema is thoroughly *un*Darwinian, as Darwin (1936, 1965) sees nothing anthropocentric in the attribution of consciousness, intelligence and emotions to animals, and emphasises that the difference is one of degree and not kind. Whilst Regan (1983) concedes that it is highly improbable that animals are autonomous in the Kantian sense, they are endowed with what he terms preference autonomy, deriving from their possession of very real preferences in the first instance, and their abilities to act so as to satisfy those preferences.

The notion that instinct is appropriate for explication of animal behaviour, but infelicitous for its human counterpart, is both pervasive and mistaken (Radner and Radner, 1996; Senchuk, 1991):

*Instinct* and *intelligence* are not parallel terms. *Instinct* covers not just knowing how to do things, but knowing what to do. It concerns ends as well as means. It is the term used for innate tastes and desires, without which we would grind to a halt. With closed instincts, desire and technique go together...But as you go up the evolutionary scale, much wider possibilities open. The more adaptable a creature is, the more directions it can go in. So it has more, not less, need for definite tastes to guide it. *What replaces closed instincts, therefore, is not just cleverness, but strong, innate, general desires and interests*...Just in proportion as automatic skills drop off at the higher levels of evolution, innately determined general desires become more necessary.

(Midgley, 1996a, pp. 332–3)

Likewise Rowlands (2008, p. 39) argues that though 'Their essence may constrain their existence...it does not fix or determine it.'

Even in the human sphere, it seems we are moved to cherish, love and protect infants and children less by rationality and more by instinctual responses; in point of fact, to the extent that we ask ourselves *why* we should so do, it is almost invariably asked after the event, as an ancillary motivation (this is not to gainsay the fact that rationality assuredly builds upon and expands our compass of concern). Likewise, the behaviour of infants towards one another is marked more by instinctual responses, but is none the less moral for it being thus; in truth, 'We are moral because we are mammalian, long before we are also rational enough to reconsider our roots' (Clark, 1997, p. 106).

Unless one subscribes to the belief that human beings are the special creation of God, the notion that our intelligence entirely supersedes instinct and that all animals are literally enslaved to the instinctual forces beyond their ken and control, is nonsensical in evolutionary terms. It would be more apt to claim that intelligence complements and extends open instincts by experience, in much the same way that culture complements and completes nature – we are 'cultural beings by virtue of our nature' (Eagleton, 1997, p. 73). It is the structure of instincts that 'as a whole, indicates the good and bad for us' (Midgley, 1996a, p. 75), and without such a structure intelligence would be incapable of accounting for conflicting tendencies and motives. We once more return to our ability, indeed necessity, to be able to reason from fact to value, for 'we can value nothing that the facts of our nature make impossible for us' (Clark, 2000a, p. 7). Truth to tell, Clark (1983, p. 191) muses, 'Nothing is present to us as valuable except through our given natures ... If we seriously say that our given natures are not to be valued, we can no longer take anything we say seriously.'

But can behaviour that is instinctual ever be intelligent, and can it ever be moral? Well, yes, and yes; we are mistaken to assume that instinctual patterns are necessarily unintelligent (Bateson, 2000), and Darwin (1936) is of the considered opinion that animals possess a rudimentary moral sensibility. Even seemingly uniformly closed instincts are not *wholly* predetermined; in order to explicate this point, Midgley (1996a, p. 307) observes that in the process of imprinting

> there is a strong, natural, internal tendency to receive from the outside an impression of a certain kind, and to *use* it in a particular, predetermined way in one's life from then on. The details of the impression are not predetermined. They must come from outside. But the tendency to form such habits is. It is a complex and positive power.

The ever-circumspect Darwin (1936) is reluctant to provide a definition of instinct, and is at pains to remind us that animals are not in total servitude to instinctual patterns, for the greater our familiarity with animals, the less likely we will be to ascribe instinctual as opposed to rational causes for their behaviour. Such an understanding is contrary to the fact that biologists have historically assumed that only learned, not instinctive, behaviour could provide evidence of consciousness (Page, 2000); Senchuk (1991) argues that animals readily exhibit flexibility, which is the principal indicator of consciousness, whilst Bateson (1990) asserts that animal behaviour involves both choice and preference. There exists no compelling evidence for the traditional assumption that genetically programmed behaviours cannot be guided by conscious thought (Griffin, 1992), with Darwin (1936, p. 447) arguing that 'a high degree of intelligence is certainly compatible with complex instincts.'

The very notion that human beings are social animals bespeaks the centrality of social instincts, and far from threatening human dignity and values, it can be seen to underpin and validate them. The Darwinian understanding of social instincts 'is not just one more set of impulses among others, but a whole way of regarding those around us, based on sympathy, which involves imagining them as subjects like oneself, experiencing life in the same way, and not essentially different in status' (Midgley, 1984b, p. 90). In identifying the inherent relationship between social instincts and moral sensibility, Darwin

(1936, p. 481) concludes that as human beings are social animals they assuredly inherit tendencies to aid and defend their fellows, and are, in common with other animals, impelled by mutual love and sympathy, for 'The more enduring Social Instincts conquer the less persistent Instincts.'

In light of the foregoing, what are we to make of Malik's (2000, p. 232) earlier contention that failure to view human beings as radically different to animals will return us to the 'beastliness' of our 'brutish' origins, for, in essence, 'viewing beasts as more human is but the other side of viewing humans as more beastly'? Although his concern is not without foundation, for he is surely correct to express alarm as to the consequences for human beings of the diminution of the centrality of subjectivity, Malik fails either to see or concede that no subjective being ought to be viewed as a means to another's ends.

Such a principle is central to social work values and practice, and what is being argued is simply an extension so as to encompass all sentient beings, in line with Salt's (1935, p. 68) conviction that 'sympathy, guided by reason, is making it more and more impossible that we should for ever treat as mere automata fellow-beings to whom we are in fact very closely akin'. It needs emphasising that human rights or dignity are in no way disparaged or endangered by a moral consideration of animals, for 'we may take it as certain that, in the long run, as we treat our fellow beings, "the animals," so shall we treat our fellow men' (Salt, 1921, p. 156). As often as not human power over nature results in the domination by the few over the many, with nature merely the subjugating instrument (Lewis, 1946).

In common speech, 'To treat someone "as an animal", is to ignore any serious wishes that they have' (Clark, 1999, p. 1), whilst to be condemned for acting *like* an animal is 'to have abandoned cultivated manners and an awareness of one's place in the social universe' (Clark, 1985a, p. 43), supposedly beholden to transitory impulses in complete disregard of the consequences for oneself or others, thereby forfeiting entitlement to membership in the moral community. That treating someone as an animal has such obviously unwelcome connotations holds force precisely because it refers to *our* actual treatment of animals.

Malik (2000) correctly identifies the continuing tendency to treat consciousness as a ghost in the machine, and the tacit assumption that objective causes of behaviour always outweigh their subjective counterparts. This is one of the cardinal errors of genetic determinism, but culpability for the effacement of subjectivity cannot be attributed solely to what Malik (2000) terms the universal Darwinists; he levels a similar charge at cultural anthropologists who assume the sovereignty of culture and deny an intrinsic human nature. The notion that humans are infinitely malleable creatures has had wide currency, especially in the philosophies of behaviourism, existentialism and postmodernism, and underlying the existentialist rejection of human nature, and any notion of essence, is the conviction that 'Statements about essences license universal necessary truths' (Cottingham, 1983, p. 465).

In an attempt to clarify the misconceptions that surround essentialism, Eagleton (1997, pp. 97, 104) asserts that

> Essentialism in its more innocuous form is the doctrine that things are made up of certain properties, and that some of these properties are actually constitutive of them, such that if they were to be removed or radically transformed the thing in question would then become some other thing, or nothing at all...we need to know among other things which needs are essential to humanity and which are not...any social order which denies such needs can be challenged on the grounds that it is denying our humanity, which is usually a stronger argument against it than the case that it is flouting our contingent cultural conventions.

Far from being a guarantee of freedom and individuality, given that 'matter is precisely what individuates' (Eagleton, 1997, p. 48), infinite malleability entails that we are interminably condemned to cultural determinism and manipulation. It is from a common nature that ethical and political implications are derived, and a consensual understanding of what it is that constitutes the moral and social good (Iyer, 1973). Contrasting pictures of human nature have implications for the philosophy and practice of social work, and in Wilkes' (1981) view, malleability is dependent upon a rejection of any notion of a substantial self and a common humanity, whereby we are all the more likely to accord primacy to impersonal systems, and lose sight of the moral primacy and value of the individual subject, and an undervaluing of individuals and situations not conducive to change. It leads to the view that the people are cases for treatment rather than respect (Ragg, 1977), and social work has been aware, especially since the 1970s, that individuals and communities cannot be moulded in any which way against their express wishes, to a vision of what *we* want for them.

The moral point underlying an assertion of infinite malleability is, in large part, a rejection of fatalism, which is often mistakenly conflated with determinism; in essence it is representative of the ongoing disputes surrounding the fundamental metaphysical questions of free will and determinism (Stevenson, 1974). In order to clarify misunder-standings, Midgley (1996b, p. 94) notes that determinism 'is simply the modest assumption of that degree of regularity in nature which is nec-essary for science, and is as necessary for the social sciences as for the physical ones'. The fact that determinism is held to be antithetical to free will (Peile, 1993) springs, in Midgley's (1996b, pp. 94–5) view, from a superstitious and over-inflated conceptualisation of both fatalism and determinism:

> A melodramatic tendency to personify physical forces and other sci-entific entities can represent them as demons driving us, rather than humble general facts about the world, which is all they have a right to be seen as. This produces fatalism, which certainly is incompatible with free-will, since it teaches that we are helpless in the hands of these superhuman beings.

The notion that determinism allows for predictability and provides gen-eral facts about our world is quite distinct from the fatalistic belief that human beings are the mere playthings of external forces beyond their control, and as such counters genetic determinism. However, Midgley (1996b, p. 101) observes that continuity of belief in a supernatural being in the background survived the much heralded death of God, and contends that

> It constantly represents human effort as an unreal cause. It shows people as helpless pawns and puppets in the grip of all sorts of non-human entities which act as puppet-masters – Nature, Entropy, Evolution, History, personified laws and forces of all kinds (notably economic ones), and most recently the selfish gene...determinism ought to be a modest assumption about the possibility of knowledge.

Whereas social work readily embraces notions of economic and cultural determinism in its understanding of individuals and society, it is reticent to acknowledge any meaningful role for biological causes. In analysing the polar extremes of biology or culture, Rose and Rose (1982, pp. 10–11) observe that the biological determinism of sociobiology was itself a response to the new left utopianism that deemed human nature to be infinitely malleable:

> The helplessness of childhood, the existential pain of madness, the frailties of old age were all reduced to mere labels reflecting dis-parities of power. But this denial of biology is so contrary to the

actualities of personal lived experience that it renders people ideologically vulnerable to the commonsense appeal of the new biological
determinism.

The notion that we are faced with fundamentally incompatible deterministic or autonomous models of human nature (Hollis, 1977) presupposes that absolute determinism or absolute autonomy are the only
choices on offer; we are either passively biological or actively social
beings. Social work is presented with two extreme and misleading
choices, the notion that we are incapable of changing that which it is
in our power to change on the one hand, and on the other the fanciful
belief that we can change *everything* (Midgley, 1983d). A not insignificant part of the problem is the conceptualisation of inherent binary
opposites: nature or culture, instinct or intelligence, feeling or reason,
determinism or free will, animals or human beings – the duality of biological or cultural. On this model human beings and their dignity rest
upon our opposition to any natural reading of human nature, which is
nonsensical, for 'The form of life natural to a creature helps to define
what happiness is for that creature's kind, what capacities are there to
be filled, what occasions are needed for it wholly to be itself' (Clark,
1997, p. 51). This has fundamental implications for our understanding
of morality and candidates for moral considerability, for, as Clark (1982,
pp. 117–18) reminds us, 'Our morality must in the end depend not only
on what we think we are, but on what we think the world is.'

Conceptions of human nature are inherent in any discussion about
what constitutes good or otherwise for human beings (Stevenson, 1981),
and Butrym (1979, p. 41) holds it to be 'the main and most significant material with which the social worker is engaged'. Notwithstanding, there is probably no more contentious and ideologically charged
concept in the domain of the social sciences:

> The term is suspect because it does suggest cure-all explanations,
> sweeping theories that man is basically sexual, basically selfish or
> acquisitive, basically evil, or basically good. These theories try to
> account for human conduct much as a simpleminded person might
> attempt to deal with rising damp, looking for a single place where
> water is coming in, a single source of motivation.
>
> (Midgley, 1996a, pp. 57–8)

One conspicuous exception in the social sciences is Abraham Maslow
(1993), who contends that human values are biologically rooted, and
that the welfare and well-being of individual subjects, and the notion

of a good society, ought to be central concerns of biological philosophy. Precisely because human beings and animals have evolved by the same processes, of necessity biology must be able to shed some light on our understanding of ourselves (Birch, 1999).

The prevalent tendency to contrast human nature with nature generally uproots human beings from the natural world and from a serious moral consideration of other animal life. This deracinated metaphysic is borne out by the major philosophers in the Western tradition; Regan (1991) relates that Hobbes conceptualises nature as wild, treacherous and, nasty, and that it needs to be brought under human dominion; for Locke, nature attains value only as human property; Kant holds that the value of nature is extrinsic, in that its value is utterly dependent upon the human observer (indeed Kant's noumenal self, which is held to be the embodiment of all value, excellence and freedom in human beings, transcends terrestrial existence), whilst Mill conceptualises nature of value as a source of aesthetic sensibility or sublime veneration (although he specifically insists that animals matter morally).

One observes in all these thinkers an assured belief that nature has extrinsic rather than intrinsic value (Elliot, 1992); they adopt a worldview wherein it is human beings who bequeath value to the natural world, especially via the medium of rationality, and it is culture that makes nature valuable. This disconnection from the natural world, and a denial or obfuscation of our biological embeddedness, serves to misrepresent human nature, and to exclude the assignably *natural* animals. Value is invariably filtered through an exclusively human lens, and in this line of vision animals are but peripheral phenomena. Because rationality is conceived to be the moral *summum bonum*, humans are deemed to transcend the natural world and biology.

Such a conceptualisation of the relationship between the self and the world reflects what Fox (1991) terms a *discrete entity ontology*, whereby the world is pictured as being constituted by discrete entities, with a fundamental emphasis upon differences; in marked contrast a *continuity ontology*, whilst acknowledging beings to have a degree of independent existence, conceives them as essentially characterised by a state of interrelationship and connectedness. If we are to understand the relationship of the self to the world in terms of a continuity ontology, Fox (1991, p. 118) contends that it commits us to the recognition that our drive towards wholeness transcends our drive towards individuality, and 'can be realized to a considerable extent through the psychological process of identification. Identification with others means empathically entering into their joys and sorrows . . . Identification remains a relationship between two identities; it is not identity'. Therein rests the central importance of relationship within the social work process, what Perlman (1979) singles out as the *heart* of helping people.

This notion of interrelatedness is discernible in Lovelock's (1979) concept of Gaia, and Clark (1983) argues that a realisation of enduring interrelatedness arises from an awakened sensibility of kinship, and that our rightful relationship with the world entails a simultaneous taking ourselves, and the whole of which we are integrally a part of, seriously (Clark, 1984). Our own well-being is ultimately dependent upon that of the whole earth, and we possess an inherent affinity and a biological attraction for the natural world and life *per se* (Kellert and Wilson, 1993; Wilson, 1984). Not only is this the only home we have and could have, but more importantly we are not sole tenants. To live as though we are aliens (Haught, 1990; Kohak, 1984) is a fallacious and destructive metaphysical drama.

It is inane to consider humankind in isolation from its moorings in the natural world; we are neither terrestrial interlopers nor displaced beings, for in reality we dwell within 'a setting of living things and creatures to which we are attuned, and to whose music we for our part are far from deaf' (Midgley, 1986, p. 87). Not the opposite of nature, culture is 'the interface between us and the non-human world, our species' semi-permeable membrane' (Mabey, 2005, p. 23). Perhaps our inability to extend due respect and seriousness towards the natural world and living creatures is more likely the cause of our notorious inability to extend due respect and seriousness to fellow human beings.

Rather than viewing our biological nature as an unfortunate and execrable factual interloper that we ought to banish at the first possible opportunity, Maslow (1968, 1993) contends that its acknowledgement and embrace has positive metaphysical, moral and psychological implications. Far from ensuring human dignity and freedom, those very ideals are endangered by the notion that we are beings who transcend biology and the natural world.

If we understand that the values we cherish are biologically rooted, and that our dignity arises from within nature, not against or apart from it, we have nothing to fear from an acknowledgement of our animality, and animals have everything to gain. Our insistence upon ontological discontinuity not only misrepresents human nature, but it invariably effaces serious moral consideration of animals.

This chapter has located humankind as a biological, social and terrestrial species, and has argued that our dignity and worth resides therein. It has also made the case that therein lies the dignity and value of other animals, and we have a moral duty to respect and care for the natural world for the reason that it does not exist as a backdrop to, or mere plaything of, loftier human designs, but to house what Stephen Clark (1977) terms the *wider Household*. Life on our planet is marked by continuity, and differences in the wide range of attributes and qualities witnessed in human beings and other animals nevertheless remain a difference of degree rather than kind. We ought, as Midgley (1996a) argues, always concern ourselves with asking how we differ *among* rather than *from* other species, and acknowledge that human dignity requires neither claims to uniqueness, nor the denigration of other animals. We share a biological as well as moral kinship:

> Yea, too, the creatures sheltering round –
> Dumb figures, wild and tame...
> They are stuff of thy own frame.

> (Hardy, 1976, p. 447)

Accordingly, it has provided the context for a revised code of ethics, which takes cognisance of our ontological continuity and terrestrial status, and provides the conceptual framework to guide our social workers through the practice dilemmas depicted in Chapter 1, thereby equipping them (and us) with the wherewithal to know how, what and why to think about, and act towards, other species.

# 4
# Social Work and Respect for Individuals

> A just and loving gaze directed upon an individual reality. I believe this to be the characteristic and proper mark of the active moral agent.
>
> *Iris Murdoch* (1996, p. 34)

At the heart of social work's ethical and moral framework (indeed morality generally) is the ubiquitous concept of respect for persons, which is almost universally seen as providing the foundation and rationale for the practice of social work, and principle from which all other social work values are derived. It is variously seen as 'the basic value in social work... [and] a presupposition of morality' (Plant, 1970, pp. 12, 20), 'a prerequisite of any morality and, therefore, of universal relevance... the central concept of practice' (Ragg, 1980, pp. 219, 231), and 'morally basic... all other moral principles and attitudes are to be explained in terms of it' (Downie and Telfer, 1969, p. 33).

This chapter will undertake an extended examination of this concept, the importance of which resides in the necessity that social work be congruently guided by a moral framework that actually translates into practice what it theoretically purports to be an article of faith. It will be argued that this pre-eminent concept is, at least so far as orthodoxy mandates it, exceedingly problematic not only all for other animals, but for many human beings, and that the alternative concept of *respect for human beings* resolves this dilemma by arbitrarily lifting the moral drawbridge so as to effect the simultaneous inclusion of all humans and exclusion of all other species from the circle of moral considerability. It will be argued that the concept of *respect for individuals*, based upon an attention to the interests, needs, welfare and well-being of a creature, not its origin, is a far more efficacious and impartial moral principle to guide social workers and to ensure the extension of respect, irrespective of species membership.

Turning to the broader philosophical literature, *respect for persons* is deemed 'fundamental in morals' (Harris, 1968, p. 129), requiring an acceptance of the intrinsic value and worth of a person (Kendrick, 1992), entailing an active concern for the welfare of *self-conscious* individuals and a respect for their wishes (Harris, 1998), and a ubiquitous principle transcending cultures (Browne, 1995).

This much confirmed, it will be worthwhile to clarify what it is that the terms *respect* and *person* are commonly held to refer to, as part of the process of securing a foundational base for a more inclusive code of ethics. The Macquarie Dictionary (1990, pp. 1449, 1270) defines *respect* as 'to show esteem, regard, or consideration for', noting that its ancient meaning entailed *consideration*, whilst person refers to 'a human being, whether man, woman, or child...a human being as distinguished from an animal or a thing...a self-conscious or rational being...the actual self or individual personality of a human being'. Contending that *respect* suggests 'that there is an appropriate attitude towards persons which can be adopted only towards persons, never towards brutes' (Maclagan, 1960b, p. 293), Maclagan (1960a, p. 193) claims that whilst *person* can be held to refer to 'ordinary human beings in respect of their nature as self-conscious agents... [this] does not mean that we must in the end maintain that all, or only, human beings should be classed as persons'.

It seems fair to say that most people consider the terms *person* and *human being* to be interchangeable; human status is deemed to be a sufficient condition for being a person, for 'Human beings are paradigm persons' (Teichman, 1985, p. 184), and are 'the deciding mark of personhood' (Dennett, 1978, p. 267).

That human beings and animals inhabit disparate moral universes is manifestly transparent to Morris (1968, pp. 490, 493):

> When we talk of not treating a human being as a person or 'showing no respect for one as a person' what we imply by our words is a contrast between the manner in which one acceptably responds to human beings and the manner in which one responds to animals and inanimate objects...to be treated as a person is a fundamental human right belonging to all human beings by virtue of their being human.

All human beings, *independent* of their attributes, ought to be respected for their biological status.

The importance attached to the concept of *person*, characterised by Murdoch's (1988, p. 323) Father Bernard as 'the highest mode of being that we know', rests on the judgement that supreme moral considerability is dependent upon satisfying the criteria deemed essential for personhood. Much contemporary literature, Darwall (1977/78) notes, is given over to arguing that respect is unequivocally owed to all persons, although he makes a distinction between *recognition* (owed to all persons) and *appraisal* respect (persons or features manifesting excellence as persons). Characterised as a 'boundary concept', its application is alternatively prized and contested (Rudman, 1997, p. 3), and the prevalent notion that the concept embraces all human beings is by no means unanimous. For instance, whilst noting that there has never existed a univocal conception, either in philosophy or in common parlance, Danto (1967, p. 110) remarks that 'not every human being is legally a person (children and idiots are not persons), and not every legal person is a human being (a corporation is considered to be a juridical person)', and Engelhardt (1986) declares that 'Persons, not humans, are special.' Here we witness the ambiguity inherent in the concept, some arguing that all attempts at definition are ultimately prescriptive, not descriptive, in nature (Macklin, 1984), with others suggesting that the concept functions in both ways (Rudman, 1997).

In its Latin etymology, *person* does not in actuality refer to a human being, rather to a mask worn by actors in classical drama, signifying that they are 'someone who plays a significant part in the drama of life' (Clark, 1985b, p. 470). It is Wilkes' (1981) contention that *respect for persons* is a relative latecomer, replacing the ancient perspective that saw the human being as a composite of body, mind and spirit; what was esteemed above all else was the human being, not the person. Whilst the term *person* is often held to be synonymous with *human*, Smuts (1999, p. 108) suggests that it has another quite distinct meaning, referring 'to a type of interaction or relationship of some degree of intimacy', and that this latter expanded sense makes reference to 'any animal, human or nonhuman, who has the capacity to participate in personal relationships, with one another, with humans, with both'. All too often the term *person* follows, rather than guides, contemporary morality, and merely declares what we have already resolved (Midgley, 1995b), and in making the observation that the criteria for personhood are more often than not a function of other people's interests, Williams (1985, p. 137) contends that 'Certainly there is no slippery slope more perilous than that extended by a concept which is falsely supposed not to be slippery.'

Notwithstanding, in Kant's (1964) universe rationality is *the* attribute that entitles one to entry into the kingdom of ends. Kant's moral doctrine leads to 'an ethic of mutual respect and self-esteem' (Rawls, 1999, p. 225), and for him *respect* refers to the recognition of 'a worth which we did not make and cannot alter; by which we concede the otherness of others' (Midgley, 1983b, p. 96). Kant 'wants us to respect humanity because it is rational, not because it is conscious' (Midgley, 1996a, p. 46), and whilst animals may be conscious, their value rests in their being means to human ends. Kant 'does not tell us to respect whole particular tangled-up historical individuals, but to respect the universal reason in their breasts. In so far as we are rational and moral we are all the same' (Murdoch, 1997, p. 215), and conceptualises persons to the exclusion of considerations of character (Williams, 1981).

Kant is no philosophical Robinson Crusoe, for in the Western philo-sophical tradition rationality is singled out for especial commendation by thinkers as diverse as Hobbes, Mill and Locke. For Hobbes this enables us to transcend the state of nature, whilst for Mill (although acknowl-edging that we share an emotional kinship with animals) and Locke it is constitutive of a life worth living, and natural rights, respectively; it is Regan's (1991) conviction that all these theories about what it is that constitutes a human person serve to obfuscate our very embod-iedness and biological and ecological embeddedness, resulting in an exaltation of reason over and above emotion and all other non-cognitive capacities, a celebration of the objective over the subjective.

In contrast, Murdoch (1997) sees love, not reason, as encapsulating the essence of morality; what is needed is a 'loving attention to a crea-ture's particularity', for our sense of *otherness* 'is the root of love and knowledge' (Clark, 1997, p. 152). Whilst noting that Kant supposes that rationality assuredly discloses our sense of *otherness*, Cordner (2002, p. 138) avers that 'the basic forms of love of other human beings already implicate us in awe and reverence. This is not because they direct us to another's Rationality, however, but because of their interdependence with an individualizing sense of another'. An awareness of the essential individuality of others entails an *I–Thou* relationship in the stead of an *I–It* variety (Buber, 1970), and Buber conceptualises such a relationship as mutually respectful (Kohak, 1984) and expressly inclusive of human beings *and* animals (Linzey and Cohn-Sherbok, 1997).

Respect, and affection, are more often than not dependent upon *unlikeness* (Greene, 1982), and Cordner's (2002, p. 6) depiction of adolescent love serves to reiterate this point:

> The shock of this experience of another is also the shock of realizing another person as absolutely and ungraspably *other*. The sense of the whole world as suddenly transformed...is a sense of having been

jolted out of oneself by a reality one cannot possess but only answer to. It is as if the centre of gravity of the world has shifted elsewhere.

A respectful and loving attentiveness to the individuality and particularity of others is indispensable to efficacious and ethical social work practice.

In the Western philosophical tradition it is almost invariably taken for granted that animals, whatever attributes they may be held to possess, nevertheless uniformly fail to meet the criteria of personhood. In a very real sense, they are conceptually and metaphysically mapped as *things*, beings to whom at best we owe gentle usage, but nevertheless means to human ends. In a legal sense, animals are property, and property cannot acquire personhood, which is belied by the fact that corporations are seen as legal persons (Francione, 1995a; Kelch, 1998). It is not possible to exaggerate the profound and pervasive influence that the exaltation of rationality, and in particular the Kantian paradigm of moral value, has had on subsequent Western thought, and the social contract tradition has likewise validated the contracting, autonomous and rational adult human being as being both constitutive and exhaustive of moral considerability, with negative consequences for disabled and disadvantaged humans, and all other animals (Nussbaum, 2006).

In making the observation that our central intellectual tradition formed its foundational views on the basis of a 'crude, extreme, unshaded dichotomy between man and beast', Midgley (1985, pp. 59–60) contends that such a metaphysic still largely holds sway; even our portrayal of aliens is informative in this context, for

> Science fiction, though sometimes helpful, has far too often sidetracked the problem by making its aliens just scientists with green antennae, beings whose 'intelligence' is of a kind to be accepted instantly at the Massachusetts Institute of Technology – only, of course, a little greater. Since neither dolphins nor gorillas write doctoral theses, this would still let us out as far as terrestrial non-human creatures were concerned. 'Persons' and their appropriate rights could still go on being defined in terms of this sort of intelligence, and we could quietly continue to poison the pigeons in the park any time we felt like it.

The characteristics designated by Quinton (1973) as being constitutive of personhood are rationality, consciousness, moral agency, as well as the capacities for moral judgement and interpersonal relationships, whilst Dennett (1978) specifies rationality, consciousness, reciprocity, verbal communication and self-consciousness. Significantly, these attributes are not shared by all humans, and many animals have the capacity to satisfy at least some of these requirements. As Pluhar (1995, p. 57) relates, 'No characteristic has yet been found that is *wholly* lacking in nonhumans and wholly present in humans', and besides

> What makes creatures our fellow beings, entitled to basic considera-
> tion, is surely not intellectual capacity but emotional fellowship. And
> if we ask what powers can justify a higher claim, bringing some crea-
> tures nearer to the degree of consideration which is due humans,
> those that seem to be most relevant are sensibility, social and emo-
> tional complexity of the kind which is expressed by the formation of
> deep, subtle lasting relationships.
>
> (Midgley, 1985, p. 60)

Given that there is no definitive proof that any being *is* a person, Schwartz (1982) argues that circumspection would be the more morally prudent position to adopt. Our ethics, Clark (1995c) maintains, rest as much upon sentiment and personal attachment as they do upon reason, for we are naturally bond-forming social creatures, not disem-bodied intellects, and our moral framework must respect and reflect this reality.

Stipulative definitions of personhood have significant effect upon our moral, metaphysical and political thinking (Teichman, 1985), for 'The classes to which we belong are not merely descriptive, but normative' (Clark, 1995c, p. 320). Some will go to extraordinary lengths in order to exclude other animals from the moral universe, even at the expense of certain categories of human beings, those deemed to be *marginal*. For instance, Dennett (1978, p. 267) insists that what is of critical import is 'not that we are of the same biological species, but that we are both per-sons...For instance, infant human beings, mentally defective human beings, and human beings declared insane by licensed psychiatrists are denied personhood, or at any rate crucial elements of personhood.' Moral agency ought not to be characterised as exhausting moral con-siderability, as the *strong personhood view*, defined by Pluhar (1987, p. 24) as entailing that 'all and only persons are morally considerable beings', would have us believe.

The notion that *only* the lives of *persons* are ultimately worthy of being lived, and 'that we should treat creatures better the more that they are "persons" ', is characterised by Clark (2000a, pp. 198, 192) as *personism*;

not an insignificant number of moral thinkers, Clark remarks, contend that we treat non-rational or non-personal human beings as animals, in order that 'the respectably *human* creatures can be distinguished from mere human beings, creatures who are our kin'. The latent perils inherent in this moral apartheid, 'our readiness to think that those unlike ourselves, the poor, the weak, the stupid, have no title to their lives' (Stephen Clark, 1977, p. 140) ought to be readily apparent at a moment's reflection; in our conceit and arrogance we fail to imagine that other lives, be they different in many ways, may still be valued by the subjects irrespective of their supposed failure to attain our lofty perfectibility. Seemingly merely a variation of sorts upon the commandment in *Animal Farm* (Orwell, 1993) that whilst all animals are equal, some are *more* equal than others, it belies the humanist creed that we all stand as equals before one another.

But not all thinkers are so eager to establish a rational elite or an oligarchy of persons, whereby there exist gradations of value that might as well be chasms in their immediate moral import, and seek to constrict the moral community by rescuing human moral patients whilst excluding their non-human fellows. Parfit (1976) argues that even if it were the case that *possible* people possess neither rights nor interests, we ought to act often as *if* they in fact do, whilst Townsend (1979, p. 93) recommends that it is 'better to extend moral concern too far than not far enough'. Nelson (1988) argues that *marginal* humans, unlike animals, have suffered a *tragic harm*, and are consequently worse off, and it is this tragedy that gives them moral priority, whilst Scarlett (1997) claims that we are all vulnerable to disabilities that frustrate human purposes and relationships, and that our sympathy for *marginal* humans grounds the moral uniqueness of the human animal. All these thinkers assume it obvious that their widening of criteria for moral considerability unequivocally excludes all other animals on the grounds of species membership, and their welcome acknowledgement of our shared vulnerabilities and consequent dependencies ignores the reality that these derive in part from our shared animal nature (Macintyre, 1999), for animals are 'our fellow brethren in pain, disease, and suffering' (Darwin, in Phipps, 2002, p. 36).

We have seen in the practice examples in Chapter 1 that this is how social workers invariably conceptualise and resolve moral dilemmas involving humans and animals; for instance, the tragedy of child abuse and neglect is without exception given exclusive consideration over the interests, welfare and well-being of animals, and we ascribe basic rights to our *marginal* kind for metaphysical and speciesist reasons. But respect for all our human kind does not require, seesaw fashion, or warrant, the moral exclusion of other animals.

In our everyday lives, we do not refer to the paradigmatic model of personhood to determine whether we ought to extend respect to those souls we happen upon. We would rightly be considered odd by the proud, beaming parents of babies Thomas, Jude, Immogen or Mirabehn were we to convey that we thought that they might, in due course, grow up to be *persons*. In spite of their rudimentary attributes (which all parents naturally hope will, in due course of maturation, flourish), one imagines that the parents consider their children *already* to be ends in themselves.

The parents, and many more besides, have good reason to be affronted by the suggestion that the children in question are but *potential* persons, that we extend respect, not so much for whom the children *are*, but for that which they will *become*, or that we only extend respect to them due to the interests of the parents. We value and respect the children because they have interests in what befalls them, and can bode good or ill for their well-being and welfare. The notion that children are but *potential* or *part* persons serves only to establish their potential or *part* moral rights, leading Melden (1988, p. 70) to assert that 'the moral status of the infant is no more to be tied to its particular condition during the human being's infancy than the moral status of one who is asleep or unconscious is tied to its condition at the time it is asleep or unconscious'.

The very notion that children are in effect second-rate humans is belied by the fact that we routinely cherish and accord them moral priority, confirming that we in actuality consider them to be of inestimable worth. Likewise, it is not anthropomorphic to observe that all animals, if we but take the time and keep an open mind, have distinct individualities, and obvious interests and preferences. Neither they, or we, are indifferent to what befalls them, and those of us who have the good fortune to share our lives and homes with animals find dogmatically blanket assertions that Jayke the cat or Tessa the dog are merely *things* not only insensitive, but patently nonsensical. Were they mere *things*, we would be advised to water, feed and exercise them in much the same way that we add water and fuel to and service the car we drive, and to feel that we have therein fulfilled our obligations, such as they are.

That we respond to animals as specific individuals belies the notion that they are merely receptacles of cathood or doghood (Hull, 1978). We take efforts to familiarise ourselves with each animal's character and personality, and we respond to each animal, not as though it were a

Platonic Form, but as a unique individual. Ethological studies have also confirmed that animals often recognise their associates as individuals, and treat them as such (Griffin, 1984), and we routinely enter into deep and abiding relationships with other animals, relationships nurtured and sustained by emotional fellowship (Midgley, 1985; Sharpe, 2005).

Sapontzis (1987) maintains that everyday moral practice identifies *interests*, not rationality, as constitutive of moral considerability, and Sumner (1987) argues that all beings with interests ought to be extended those *prima facie* rights appropriate to them, for 'maximum moral respect is due to any being, human or nonhuman, who is capable of caring about what befalls him or her' (Pluhar, 1995, p. xiii). In run-of-the-mill practice, social workers attend first and foremost to the interests of those with whom they work, and in the practice examples provided in Chapter 1 the animals are due our concern because they care about what happens to them, and have interests and welfare that must be attended to and taken into account for *their* sakes.

One cannot help but wonder whether it is the dependency of babies, children, the demented and the senile, and the insane, those whom Downie and Telfer (1980) refer to as *sub-normal humans*, as much as their diminished rationality and autonomy, that is a significant factor in our assignation of lesser value. Contemporarily, we are urged to avoid dependency like the plague; couples are disparagingly said to be *co-dependent*, and we take it as a gauge of maturity that individuals stand alone. Whilst paying lip service to the importance of extended family, we nevertheless tell ourselves that Gran or Cha would much rather be in the nursing home than surrounded by loved ones, for *they* do not want to be a burden on anyone, and that they much prefer that their every material need is attended to than the familiar affections, traditions and rhythms of place. By way of contrast, many traditional cultures conceive dependency as central to our understanding of society and the common good (MacIntyre, 1999), a weakening of ties and roots leaving us vulnerable to despair (Weil, 2002).

The belief that children ought to avoid dependency is highlighted by Greer (1971, p. 236), who advocates an 'organic family' whose point it is 'to release the children from the disadvantages of being the extensions of their parents so that they can belong primarily to themselves. They may accept the performances that adults perform for them naturally, without establishing dependencies'. Any concerns raised as to possible deleterious effects wrought by child care are automatically construed as finding fault with the parents, and we are urged to swallow whole the doctrine that children invariably thrive in care, notwithstanding *any* evidence to the contrary (Biddulph, 2006; Bowlby, 2005). In spite of the fact that we are naturally bond-forming creatures, we are duly informed that all that children require is quality time.

None of the foregoing ought to be construed as endorsing that responsibility for caring for children, or for our elderly, should fall over-whelmingly upon one or both biological parents, or offspring. The very notion that the bond between mother and child is natural has had the unfortunate effect of denying any moral credit to the woman upon whom falls the responsibility for raising her children on her own (Clark, 1999), but this said, few people sincerely believe that an institutional upbringing is preferable to that which a family can offer (Midgley and Hughes, 1983).

This dread of dependency is likewise reflected in contemporary hos-tility to the concept of welfare and welfare recipients (Marston and Watts, 2004; Peel, 2003) – even some welfare agencies eschew the ter-minology – and critiques that welfare invariably fosters a demoralising culture of dependency (O'Connor, 2001), best redressed by the clarion call of individualism and free market ideology (Saunders, 2004). Both our biological and social dependency belies the contemporary cult of the splendidly detached and self-sufficient individual, and the disavowal of the moral responsibility of the state for the maintenance of conditions that are requisite for the nurturing and blossoming of human faculties (Green, 1986; Tawney, 1938). Ethics and dependency are natural corre-lates, for the very foundation of *all* morality is responsibility for others (Bauman, 2001).

A less stipulative and exclusive formulation of the attributes or char-acteristics that qualify a being for the status of *personhood* is provided by Feinberg (1986, p. 262):

> In the commonsense way of thinking, persons are those beings who, among other things, are conscious, have a concept and awareness of themselves, are capable of experiencing emotions, can reason and acquire understanding, can plan ahead, can act on their plans, and can feel pleasure and pain.

This posits a not dissimilar notion to Regan's (1983) *subject-of-a-life* criterion; both transcend the paradigmatic conceptualisation as the cornerstone and coinage of personhood and moral status, and both specifically deny species exclusivity.

We have already noted the tendency of *personists* to dismiss any-thing other than rational perfection, but a community constricted in accordance with species membership is often adhered to just as tenaciously; 'every human being, however immature or defective, who has any mental capacity at all, is a person and worthy of respect' (Harris, 1968, p. 129), and 'every human being, whatever their distinctive characteristics or lack of them, is precious and irreplaceable . . . unique in a way that nothing else in nature is' (Gaita, 2002, pp. 165, 78 – in contrast, Gaita characterises animals as possessing an attenuated individuality and hence replaceability). But such certainty that only human beings possess inherent value and merit absolute respect, and that species membership *a priori*, and uniformly, excludes all other animals irrespective of any significant attributes and characteristics, is puzzling.

Sapontzis (1987, p. 52) argues that in part the confusion derives from a failure to make a distinction between *metaphysical* and *moral* conceptions of personhood, the functions of which are to *describe* and to *evaluate* respectively:

|  | Metaphysical Person | Moral Person |
|---|---|---|
| Function | Describes a certain kind of thing | Assigns a certain moral status |
| Content | Denotes all and only human beings | Denotes creatures with rights |
| Contrast | Separates persons from inanimate objects, machines, plants, animals and spirits | Separates persons from nature and property |

A moral conceptualisation of personhood does not *a priori* exclude non-human animals from moral considerability, and the belief that biological status determines personhood, and that biological kinship confers the extension of respect to otherwise human non-persons is belied by the fact that humans who don't conform to the physiological norm, for example those with gross deformities and foetuses, are not deemed to be persons in a metaphysical sense unless or until they *look* human (Sapontzis, 1987).

We are faced, Clark (2000a) believes, with a combination of *personism* and a residual love of kin; on the one hand we accord preference to a rational elect, whilst on the other we deign consideration to our kin because they are of our kind, whilst simultaneously banishing animal non-persons irrespective of the fact that they possess similar attributes to human non-persons. In a form of moral eugenics, we sift the morally considerable wheat from the marginal chaff, whilst often continuing to pay lip service to catch phrases such as the *dignity* and the *sacredness* of human life:

If people are to be given more or less protection because their lives are judged to be more or less worth living, it is difficult not to suspect that they are valuable only for what they produce. If it is reasonable to kill a child to spare it pain, because its life can never be one that the judges think worth living, why is it reasonable to sustain such lives, at public expense once they are being lived? If abortion or infanticide of the 'disabled' is permissible, can their own later judgement that they choose to live, be granted any reasonable weight? Personists reply that *later* disabled people have their wills, and judgement: earlier they have no wills at all, and therefore are not frustrated. Only those who know what they would be missing have a right, or a capacity, to claim their lives – but no one believes them when they say that lives *like* theirs are worth preserving.

(Clark, 2000a, p. 270)

The circle of moral considerability thus constricted leaves no room at the inn for the less than perfect; the stripped, beaten and left-for-dead man taken pity upon and extended loving-kindness by the Good Samaritan ought to have been grateful that his rescuer did not question either his equality, or postpone his solicitude, until such time as he could confirm that he was a *person*. Furthermore, those who defend such moral bifurcation need to articulate the nature of the supposed morally relevant differences that justify dissimilar treatment for essentially similar creatures (Pluhar, 1987).

Having provided an overview of the philosophical rendering of the concept of personhood in the Western tradition, we shall now turn our attention specifically to its conceptualisation in social work literature, where it is almost invariably understood evaluatively rather than descriptively (Clark and Asquith, 1985; Downie and Telfer, 1969).

Downie and Telfer (1969, p. 29) state that the 'formal object of respect is "that which is thought valuable" ', and their utilisation of the notion of *active sympathy* is derived from Maclagan (1960a, p. 211), who differentiates between what he terms *animal sympathy*, 'consisting in a sort of psychological infection of one creature by another...[with] little or no sense of others as independent individual centres of experience', *aesthetic* or *passive human sympathy*, 'a more distinctively human mode, to which the consciousness, or representation, of others as experiencing subjects is essential', and *active human sympathy*, 'the sympathy of practical *concern for* others as distinguished from simply feeling with them'. In Maclagan's (1960a, p. 212) view, *passive sympathy* forms the natural matrix of *active sympathy*, the latter standing 'at the very threshold of Agape'.

In stating that *respect for persons* 'resembles the basis of rules or the reason for helping to achieve human purposes rather than a rule or a purpose', Timms (1983, p. 59) declares that it has direct reference to action, for 'It extols a range of attitudes, or readiness to act towards people described in morally preferred ways.' Concurring, Downie and Telfer (1969) maintain that *respect for persons* is a principle for action, ultimately explicable in terms of an *attitude*, whilst Clark and Asquith (1985) argue that the right to respect is an absolute universal right, committing social work to a deontological morality. We shall come to observe that social work literature's treatment of the concept of *respect for persons* evidences an ongoing tension between the priority of reason and the virtue of love.

Respect for the uniqueness of each human being is a ubiquitous article of faith in social work literature, 'irrespective of race, colour, creed or any other contingent attribute' (Clark and Asquith, 1985, p. 47), and is not dependent upon attributes, behaviour or social roles, all of which Horne (1987, p. 12) asserts are 'morally arbitrary'. We respect human beings *because* 'of what it is to be human...a precious and irreplaceable individual' (Gray and Stofberg, 2000, pp. 58, 59), and our common humanity 'presupposes an equality that admits of no distinctions between human beings of equal value' (Wilkes, 1981, p. 68). Thus *respect for persons* based upon compassion – literally a 'suffering with...a name for this realization of another' (Cordner, 2002, p. 81) – as opposed to rationality, holds that human beings are owed respect because they are alive and have interests (Tilley, in Ragg, 1977). Given the foregoing, it is unclear as to why respect is seen as being axiomatically constricted to human beings, given that species membership is itself a morally arbitrary property (Rowlands, 1998).

But the inclusive notion of *respect for persons* we have thus far encountered is not a uniform view within the literature. Whilst suggesting that the attitude of respect may best be likened to the Gospel language of *agape* or *caritas*, Downie and Telfer (1969, p. 29) proceed to claim that respect or agape are fittingly directed towards persons, 'conceived as rational wills...an attitude which combines a regard for others as rule-following with an active sympathy with them in their pursuit of ends', which it must be said is a travesty of the Biblical sense of love for the individual (1 *Corinthians* 13); Jesus' (*John* 15: 12) injunction to 'love one another as I have loved you', does not come with a disclaimer that we ought to fittingly direct our love towards *only* rational wills. Rather,

'What matters is not to think much, but to love much' (St. Teresa, in Mascaro, 1970, p. 36), and the notion that we ought to respect only rational wills is rightly seen, in the context of our everyday lives, as a nonsense.

It ought to be obvious, by such stringent criteria, that many humans do not possess the designated prerequisite capacities, and Downie and Telfer (1980) maintain that we ought to consider such people as having an entitlement to an attenuated rather than normal respect. We are surely mistaken to insist that intellectual capacity be the prerequisite of basic moral consideration, rather, it is a being's capacity for emotional fellowship, evidenced in sensibility, as well as social and emotional complexity, enabling them to form profound and enduring relationships, that is central, and this we share with other animals (Masson, 2003; Midgley, 1985; Sharpe, 2005).

Whilst maintaining that the value of the human person, deriving from their capacity to experience emotion, is in no way inconsistent with the exercise of rationality, 'for in so far as emotions are characteristically human they necessarily involve rational will', Downie and Telfer (1969, p. 22) posit that animal emotion is of another and non-rational order – 'It is true that some animals may be able to experience certain emotions, but the ability to feel and express a wide range of sustained emotions is characteristically human, and it involves the perception and discrimination which only reason can supply.'

Whereas reason is usually contrasted with feeling, and human beings and other animals characterised respectively, Midgley (1996a, pp. 256, 262) comments upon the perversity of equating rationality with cleverness and mere intellect, for rationality is 'a priority system based on feeling', and not peculiar to animals of our kind:

> There are, I think, two distinct elements in rationality: cleverness and integration. By integration I mean having a character, acting as a whole, having a firm and effective priority system. The second is a condition of the first, not the other way round. For the full respect that we give to rationality, we need both. But integration alone is something of enormous value, and respect seems a suitable name for the recognition with which we salute. And integration is not confined to people.
>
> (Midgley, 1996a, pp. 256, 262)

In remarking that *respect for persons* is a uniquely absolute and universal right, Clark and Asquith (1985) observe that the values and rights derivative of this principle are qualified or limited. *Respect for persons* is perfectly compatible with the fact that an individual's capacity for exercising autonomy and self-determination cannot always reign supreme (Bernstein, 1975), but ought always to be seen within a context, and at times must be subservient to a person's longer-term interests (Horne, 1987). We might on occasion decide that an individual foregoes their right to confidentiality or self-determination, for instance, when disclosing suicidal or violent intent. We rightly consider that no individual has an absolute right to treat others or themselves in whatever manner they so desire, and that in so doing we are not invalidating the principle of respect.

The important point the principle of self-determination seeks to make is that, all thing being equal, we ought to respect and indeed value an individual's right to make decisions about *their* life. In this sense it can be categorised as a negative freedom, a right to non-interference (Berlin, 1986), a motivation to avoid doing harm rather than a desire to do good (Wilkes, 1981), for 'Negative duties, not directly to cause evil, are more universal and more powerful than the positive ones, to prevent evil or cause good' (Clark, 1997, p. 163). Indeed, Miller (1968) insists that respect dictates that social work ought not to concern itself with imposition upon involuntary clients, rather with advocacy upon behalf of those who seek out its services.

Those who link self-determination with positive freedom 'tend to play down its status as a right, and to emphasize its role as an ideal or end to be pursued in the casework process' (McDermott, 1975, p. 7), as witnessed by Bernstein's (1975, p. 40) conviction that human worth 'is based only moderately on what people are; much more on what they can be'. Utilitarian in its emphasis upon consequences, and an ideal that dominates social work thinking, it lends itself to the manipulation of those with whom social workers work. It surely had to follow that not only would *clients/consumers/customers/users* be so treated, but social workers also, with the prevailing emphasis upon *teamwork*, or what Wilkes (1985) depicts as an ideology of *belongingness*, entailing a diminution of respect for the individuals who comprise the team (Sennett, 1999), for 'individuals justify an institution more than the institution justifies its members' (Friedlander, 1994, p. 99).

This said, negative freedom is also surely limited, for what I or others do, or who we are, cannot be inconsequential concerns. If we are not to lose our moral and social bearings in our everyday world, it is essential that we receive the natural and sincere responses of those about us (Midgley, 1995b). Whereas positive freedom assumes that human nature is mouldable and conceptualises value as dependent upon the possession of particular attributes, Wilkes (1981) claims that negative freedom posits an essential and unchanging human nature, and that a human being's basic dignity, independent of particular attributes, derives from something *inherent* in their being. The greater an individual's capacity for agency, the greater their right to non-interference (Pluhar, 1995), but this does not endorse a two-tiered moral status between them and moral patients.

Historically, social work has bestowed pride of place upon positive freedom, but a commitment to social justice must not be pursued at the expense of respect for the individual, for what he or she *is* (and often notwithstanding what he or she *does*); not infrequently the professed article of faith that humans being are ends in themselves is supplanted by a commitment to effecting change or adjustment (Wilkes, 1985), experienced by those on the receiving end as petty tyranny and bureaucratic regimentation (Chesterton, 1912). The fancy that we can bestow upon ourselves the omniscience to mould others to socially engineered ends, or towards an ideal, leads all too readily to hubris and contempt, respectively. Truth to tell, precious few of us are, or will ever be, the finished article, and why we ought to cease to respect an individual once we deem that they are acting in any manner that bespeaks an absence or lack of reason is to treat the principle of *respect for persons* instrumentally.

In claiming that the principle of *respect for persons* 'provides the "means" by which the social worker creates and presents a picture of the "subjective" characteristics of the client as an individual', Horne (1987, pp. 94–5) maintains that values in social work 'are basically instrumental to the purpose, and appear no more than a "means to an end" ... the "end" rather than any moral obligation is the justification for the act'. By way of contrast, both Halmos (1966) and Wilkes (1981) contend that values cannot be conceived as instrumental, but are derived from either a moral imperative or some ultimate conviction. If *respect for persons* is conceptualised deontologically as a basic moral right – by definition 'the ground of a moral obligation ... not a consequence of our having a moral obligation' (Regan, 1982, p. 117) – respect is owed because of who the individual *is*, and not dependent upon consequences, or what he or she may *be*.

Subjectivity, as depicted by Horne (1987), Howe (1979) and Philp (1979), has a contingent status. Indeed, Philp (1979, p. 98) contends that social work 'cannot make people when an individual's act has removed him from the right to be perceived as human', whilst Horne (1987, p. 99) argues the creation of the subject 'is also limited as it exists only within the boundaries of the extent to which society sanctions it'. Their acknowledgement of the significance of subjectivity is diminished by their rather mystifying claims that social work somehow makes people or creates subjects, as though it were in *our* power to do so. To envisage subjectivity as something *created* rather than *attended* to inevitably favours an instrumentalist morality that fails to respect the irreducible worth of the individual subject. The further argument that one *becomes* a person, essentially via a social process, and that 'a human creature reared in isolation from human contact...could not conceivably be a person' (Clark and Asquith, 1985, p. 17) ignores the fact that moral considerability holds regardless of our relationship to others (Pluhar, 1995); whilst such a creature would undoubtedly be stunted in so many ways (for we are, unquestionably, a social species), he or she is no *tabula rasa*. What is of value in us can be neither culturally nor historically relative, but must correspond to the needs and range of our nature (the same holding true for all other species). Our value, as is that of all subjective beings, is inherent, not acquired.

We do not create subjectivity (indeed subjectivity is what it is, and that *is* the point) any more than we create individuals; to attempt to do so entails a retreat from, rather than engagement with, subjectivity. The fact that an individual's subjectivity may be overwhelmed by their objective status, in the eyes of society at least (a point not being contested), to the extent that they are no longer seen as human (the point being specifically challenged), is surely all the more reason that subjectivity be regarded as substantial, transcending cultural composition or ascription, something encountered by loving attention. *Pace* Philp (1979), social workers often are the ones left to convey back to the wider society the common humanity of those whom society has deemed to have forfeited their standing because of their behaviours, working alongside those making attempts at moral redress or redemption.

Respect, Butler (1886, p. 425) posits, is owed to those who have interests, for 'When we rejoice in the prosperity of others, and compassionate their distresses, we, as it were, substitute them for ourselves, their interest for our own.' As Watson (1978, p. 36) observes,

> On Butler's notion of compassion, as substituting others' *interests* for one's own, an attitude of compassion towards human beings entails respect for them as creatures with interests. It does not entail respect for them as creatures with the ability to adopt rules which are held to be binding on oneself and all rational beings.

This position represents the key to the code of ethics to be articulated in the Appendix, that we ought to attend to *all* creatures with interests, irrespective of their rationality or capacity for moral agency; in practice, social workers do not set the bar so as to sequester moral agents as solely morally considerable, to the exclusion of all others. The latter may conceivably fall well short of the paradigmatic model of personhood, but if we take their interests into account, suddenly we become aware that the moral landscape is inhabited by myriad other creatures, human and non-human alike, and we are struck by the inadequacy of a model of autonomous and contracting agents to ground our duties and obligations to our fellow creatures. Consequently, the case for a revised code of ethics will be developed in light of this reality, and is underpinned by Regan's (1983, p. 171) *subject-of-a-life* criterion, whereby 'A *sufficient* condition of being owed such duties (of justice) is that one have a welfare – that one be the experiencing subject of a life that fares well or ill for one as an individual – independently of whether one also has a conception of what this is.'

We will now turn our attention to the ingenious and somewhat disingenuous attempts to rescue *marginal* humans whilst at the same time excluding all other animals. Those adamant that only *full* persons are possessed of rights have their work cut out to justify preferential or differential treatment of their supposedly *marginal* fellows from that shown towards other animals (Dombrowski, 1997). Interestingly enough, proponents of this position make recourse to a characteristic or attribute that they initially and ostensibly disavow. For instance, Clark and Asquith (1985, p. 16) argue that personhood is not a list of required attributes but ultimately ascribed by social processes, and whilst personhood is not independent of physiological and biological attributes or sentiency, 'A [biologically] human individual is not automatically a person.' In their view, the aforementioned characteristics fail to take account of essential and uniquely human attributes such as consciousness and rationality:

> if we accept the determinist and in particular the biological and physiological position, there is prima facie little reason to respond to human beings whether born or unborn in a manner different from our reaction to other biological organisms. Yet in terms of our ordinary commonsense morality we do conceive of human beings as different from animals.
>
> (Clark and Asquith, 1985, p. 13)

It would appear that Clark and Asquith consider the inclusion of animals in the moral circle as antithetical to human dignity and uniqueness, for they are merely determined creatures. We cannot make recourse to the argument that species membership matters in any absolute sense (*especially* if we argue that a biological human is not automatically a person), as distinct from the notion that species membership is an important consideration. The belief that kinship is a sufficient justification for the inclusion of *all* human beings and the exclusion of *all* other animals is profoundly misguided:

> Kinship, broadly construed, warrants preferential treatment of one being in comparison to another because we have *acquired duties* to one and not the other. However, one can only have duties, acquired or unacquired, to beings who are *already morally considerable*... We construe our duty to respect others' lives as a 'natural' or *unacquired* duty, holding regardless of our relation to those others.
>
> (Pluhar, 1995, p. 166)

Our tendency to be partial to our own kind, especially to close kin, is neither unexpected nor extraordinary, but these tendencies are not in any way absolute. Indeed they are remarkably flexible and species non-specific, and provide no justification for indifference, for 'Kinship interpreted in terms of closeness can be used to justify the favoring of one morally considerable being over another, without violating the other's basic rights, but it cannot be used to show that a being *is* morally considerable' (Pluhar, 1995, p. 166). It does not do to acknowledge our biological status, but then adhere to an ontological creed of fundamental discontinuity.

Whilst Downie and Telfer (1969) acknowledge that animals can be said to possess personality in a minimal sense and to have a degree of sentience in common with human beings (indeed they contend that sentience in human beings is the foundation of self-determination and rule-following), no normative implications are held to follow apart from 'a duty to avoid causing them unnecessary suffering'; but, as Stephen Clark (1977, p. 44) observes, 'It is of little use claiming that it is wrong to inflict unnecessary suffering if anything at all will do as a context for calculating necessity.'

Clark and Asquith (1985) suggest that if we were to take sentience as an essential attribute of personhood, then we would have to look upon abortion and the slaughter of animals as constituting murder, as though the dismissal of such arguments is warranted solely on the grounds that they offend or are contrary to contemporary moral sensibilities. As history shows, the circle of morality has consistently been widened as a result of recognition of inconsistencies and an awareness of the arbitrary nature of much of our prior moral thinking. We can choose either to pull up the drawbridge, or to follow where our considered reflections may lead us:

> By selecting different words for what others might call a victim, we are free of blame: call a thing a neonate, an embryo, a pre-embryo; call it an oncomouse, an animal preparation or a walking larder. Those particular evasions are aided by a curious piece of doublethink: the preferred expressions are chosen as being devoid of any moral force of a contentious kind, but then employed to justify what would have been contentious morals. First we insist that the moral question must not be begged, and so rule out such words as 'murder' in favour of 'homicide' or 'termination'; then we infer that since the act, so described, lacked any moral import, we may properly perform it. If things are only and entirely what 'we' call them, argument becomes irrefutable, and hence impossible.
>
> (Clark, 1998d, p. 28)

Clark and Asquith (1985, p. 20) contend that even those human beings who lack the traditional prerequisites of personhood can nevertheless be accorded respect because of species membership – 'once an entity is established as the possessor of the moral status of personhood it is entitled to treatment on certain principles, even if its possession of the usual attributes of personhood is questionable or incomplete'. To argue against specific attributes for human beings, whilst simultaneously excluding animals because they supposedly lack those very same attributes, is to apply moral double standards. It is no argument to dismiss the interests of animals on the grounds that they are unlike us (indeed their unlikeness ought to awaken us to their independent reality) or that, as social workers, we ought to attend to more deserving cases:

> Moving from the exalted to dangerously near the banal, it is possible that personhood is extended to domestic pets. When people say their dog is one of the family they may be speaking more literally than metaphorically. The pet may well receive every possible care and comfort, denied to arguably more deserving persons away from the family orbit.
>
> (Clark and Asquith, 1985, p. 18)

Having made the observation that animals are indeed often considered as kin, Clark and Asquith (1985) proceed to place the shackles upon any widening of the circle of moral considerability via an enjoinment against straying from tradition (which is a little odd given their argument that the concept of personhood is culturally and historically relative). The implication is that solicitude for animals is at best misguided (notwithstanding *possible* domestic animal personshood), it being morally preferable to direct our concern, energies and resources away from undeserving (animal) to more deserving (human) cases. Charity, it seems, does not begin at home after all.

We have seen that attempts to identify supposedly unique human attributes, and then to insist that they are discontinuous, as well as the prerequisite of moral considerability, are, from an evolutionary perspective, fundamentally misguided. Likewise, the veneration of rationality has been seen to be extremely problematic for the moral standing of many human beings. As if those hurdles were not substantial enough obstacles, Ragg (1977, p. 62) contends that language possession is the *prerequisite* of consciousness, personhood and moral agency, for 'What is outside language or language-dependent forms is unknowable.'

The logical implication of Ragg's conceptualisation bodes ill for the many human beings who are not language users (at a sufficiently abstract or coherent level prized by linguistic philosophers), who therefore cannot be conscious beings (at least in the reflective sense), and as a consequence cannot be persons (in the *full* sense so revered by moral personists). Why I would cease to be a person, should I suddenly be struck dumb, and by added misfortune lose my ability to converse via the mediums of written or sign language, is a tad mystifying to say the least; that I would also forfeit my consciousness for want of linguistic practice only compounds the incredulity. If my child never acquires the gift of speech, ought I to assume that the child is, and will always be, an unconscious creature? As Fodor (1999, p. 12) observes, 'it isn't clear how being able to talk could *create* thought or consciousness since, to put it mildly, it isn't clear how a creature that can't think and isn't conscious could learn a language.'

Ragg lowers the bar somewhat by contending that personhood is an attribute that can be *learned*, and that linguistic inability may be compensated for by gestures and non-verbal communicative expressions. Whilst acknowledging the role of non-verbal communication, Ragg (1977, p. 101) insists that the absence of language entails that 'the experience is relatively inaccessible to the caseworker; more important though, it is also inaccessible to the client'. Leaving aside the contentiousness of his latter claim (which assumes that *all* experience is language-dependent), why it is that we ought to extend consideration of non-verbal communication in our dealings with fellow humans, whilst simultaneously outlawing it in our dealings with other animals, and why it is that we ought to treat characteristically similar individuals in morally different ways, is not elucidated.

In declaring that it would be anthropomorphic to attribute consciousness to animals, Ragg (1977, p. 25) denies animals personhood because they do not engage in the highly evolved linguistic practices of human beings, observing that 'it would be fair to say that any animal that could speak would be a person'. Parrots aside, this presupposes human language to be a unique evolutionary development, but an overly abstract notion of language serves to obscure its continuity with other forms of communication (Midgley, 1996a), and its dependence upon our animal nature and inheritance (MacIntyre, 1999).

Likewise, it is anthropocentric and *un*evolutionary to suppose that animals, in the absence of human language, can never acquaint themselves with the world or reality. Animals are not blind, deaf, dumb or senseless automatons; they are intimately acquainted with their world, for, like us, their lives depend upon it. The absence of human language or speech does not preclude intentional states in human infants and many animals, for 'Only someone in the grip of a philosophical theory would deny that small babies can literally be said to want milk and that dogs want to be let out or believe that their master is at the door' (Searle, 1983). Indeed Gopnik (2009) argues that we have profoundly underestimated the consciousness of human babies, a more open-ended and imaginative awareness than that of many adults.

Expressions of concern for other beings, be they human or non-human, are dependent upon a whole raft of other considerations besides rationality or language possession (most notably consciousness, sentience and emotional fellowship), and cannot be reduced to one cardinal moral attribute that by coincidence is deemed to be an exclusively human characteristic. Even Kant does not suppose that language must be present for us to have any degree of concern in the first instance (Midgley, 1996a).

Downie and Telfer (1980, p. 40) argue for three levels of concern:

> On the lowest level are the animals, who are regarded as having a presumptive right not to suffer…Next we have what we may call 'sub-normal' humans, who are not accorded full respect but are not treated like animals either…Finally we have the normal humans

who are accorded full respect. We may describe the distinction between the sub-normal and normal human beings by employing the evaluative concept of a *person* to mark off those human beings who are worthy of full respect for the individual.

Were this to be the moral lay of the land, it would assuredly be a fact of grim portent for many of those with whom social workers work. Although rationality is ostensibly seen as the distinctive all-or-nothing attribute, on closer inspection it rapidly becomes clear that in practice there exists one rule for human beings (*sub-normals* included), and another for all other animals. Whilst animals ought to be treated with kindness – Coetzee's (1999, p. 61) Elizabeth Costello reminds us that in its full sense kindness refers to 'an acceptance that we are of one kind, one nature' – and compassion, Downie and Telfer (1980, p. 40) claim that they cannot be considered as having value in and of themselves, for there are certain ways in which animals are routinely treated that we would not tolerate were they done to *any* human being:

> Leaving aside the questions of using animals for food and for experimentation, one may simply consider the question of transplant organs. If there is a chance that an organ from an ape, say, might do for a man, a doctor will not hesitate to kill the ape. But no one would kill a mental defective for transplant material, even if it were certain that his organs could save several people.

Given that the human poor and vulnerable have historically been subjected to medical experimentation (Black, 2003; Weyers, 2007), and given subsequent developments since the publication of their book, Downie and Telfer's certitude seems patently misplaced; the 'farming' of humans for their organs (Goodwin, 2006), especially in Third World nations (Interlandi, 2009), has become a thriving trade, and if anything, in the future, what with biotechnological fetishism and the headlong and seemingly unstoppable rush towards genetic engineering, embryonic research and cloning promises more, and worse, of the same (Appleyard, 1998; Rollin, 1995; Uren, 2002), ushering in a technocracy (Clark, 1998b).

Contrary to Downie and Telfer's (1980) certitude, we ought not to consider it a virtue to sacrifice a primate without a moment's hesitation, nor take it for granted that human ends always justify the means and must invariably prevail, so that when confronted with the aforementioned transplant it seems *obvious* that this is morally permissible, and just as obvious to castigate all moral objections as undeniably misanthropic. Frey (1987) suggests that logic decrees that to the extent that we find it morally repugnant to treat *defective* humans instrumentally, then the case against treating animals in such ways is likewise strengthened.

We need to identify a *morally* relevant difference that would justify the treatment of animals in ways that would be prohibited with human subjects, and that such a requirement of necessity entails restrictions on human interests (Jamieson, 2002). Attempts to confer greater value upon the life of a *full* person than that of an ape inevitably entail that the life and rights of supposedly marginal humans are cancelled out by the needs of *full* persons (Pluhar, 1995). It is incontrovertible that the great apes manifest emotion, individuality, personality, as well as intelligence and reason, and have mental capacities and emotional lives similar and approximate to those of our own (Fossey, 1983; Goodall, 1990). The repudiation of the utilisation of animals as organ donors is not because it never leads to benefits for human beings, but rather that such gains are always ill-gotten (Regan, 1989), for they invariably fail to show respect to the individual animals involved.

*Marginal* humans are, it turns out, to be treated differently to animals because they are kin; although failing to satisfy the requirements of personhood, they

> are nevertheless given the courtesy title of "human". This extension of respect is not strictly rational, though it may be a likeable and attractive sort of irrationality, and in any case is probably ineradicable, the result of a biologically determined sense of kinship with other members of one's biological species.
>
> (Downie and Telfer, 1980, p. 48)

However, we have already seen that kinship arguments cannot be utilised to show that an individual *is* morally considerable and extending them a modicum of discretionary respect, like scraps from the moral high table, only confirms Chesterton's (1949) observation that madness is not an absence of reason, rather the exclusion of everything else. Respect ought not to be deemed supererogatory, rather indispensable to our remaining moral beings.

In a bid to ground our obligations to supposedly marginal humans, it is variously argued that we ought to view babies and children as *potential persons*, the demented and the senile as *lapsed persons*, the mentally ill as *temporarily lapsed persons*, and the insane and those with profound and irreversible mental deficiencies as *permanently potential persons* (Budgen, in Horne, 1987; Downie and Telfer, 1969). It is also Downie and Telfer's (1969) contention that there exist sufficient *resemblances* between *normal* and *sub-normal* human beings to warrant the latter's inclusion under the umbrella that is the principle of *respect for persons*, and that the attitudes of *affection* and *pity* (whilst not moral attitudes, they are in essence consistent with *agape*) also serve to ensure the extension of respect. They further argue that the notion that human beings possess equal worth is not a factual statement, rather a normative moral principle (Downie and Telfer, 1980).

It hardly needs articulating that there exists a world of difference between respect for a *person*, and respect for an *individual*; to speak of individuals in terms of their potential or lapsed personhood, or to extend respect due to resemblances, are in effect to do so indirectly. On this reading, perhaps *respect for persons* is little more than an expression of pity, and can all too readily lead to an attitude of contempt (Simpkin, 1979), for whereas pity is largely self-centred, respect is other-centred. Respect is conceptualised as *beneficence*, as something extended out of courtesy or as a favour or concession, rather than as *recipience*, a right to be treated in ways that are constitutive of respect.

It is Downie and Telfer's (1969, p. 35) conviction that the extension of the principle of *respect for persons* to human non-persons 'would not be possible... unless we first knew what it was to adopt it towards normal persons'. But to conceive normality as the ground of moral obligation and respect as its consequence is mistaken, for a basic moral right, in this case, respect, is not the consequence of our having a moral obligation, rather it is the ground of moral obligation. We ought not to treat the supposedly *marginal* with respect only on the grounds that I know what it is to respect you as a *normal* human being, and you in turn are polite enough to return the favour. Would it not be infinitely preferable to respect others for their individuality (in all its hues and manifestations), and to treat others (independently of their attributes) as we ourselves wish (or might wish, were we to lose even that faculty) to be treated (especially were *we* to be in *their* shoes), there but for the grace of God, Fate or Chance?

Given the modifications and qualifications that are endemic to discussions as to the nature of personhood, it is not in the least surprising that there are expressions of grave concern that the principle of *respect for persons* either actively promotes discrimination against those who are deemed not to be persons (Timms, 1983), or engenders discrimination against supposedly marginal humans (Watson, 1978, 1980). Centring upon the capacities of rationality and rule following, the principle of *respect for persons* must stigmatise individuals not exercising these capacities, though, so long as they are still possessed, though not exercised, the profession has some moral reason to care for these individuals... However, given that the capacities... are hardly possessed at all by some severely mentally handicapped individuals, 'respect for persons' provides little and no moral reason, respectively, for the care of these individuals. We are only obliged to control them. Caring professions working with these individuals must find another principle if talk of 'caring' is to be taken seriously.

(Watson, 1980, pp. 59–60)

Wilkes (1981, p. 3) argues that the notion of caring has been replaced in social work by caring for those deemed *important*, with undervaluing of those without specific problems, 'namely the old, the crippled, the sick, the dying, and all those who are afflicted with the sorrows of suffering and death'.

A significant part of the problem, Watson (1978, 1980) claims, is that the relevant categories of characteristics that warrant respect are unnecessarily restrictive, paralleling Warren's observation (1997) that morally relevant considerations are often conceived in a *uni-criterial* rather than *multi-criterial* manner. In place of Downie and Telfer's (1969, p. 25) interpretation that showing active sympathy for 'persons as rational wills' entails 'making their ends our own', Watson (1978, pp. 43–4), in arguing that ends need not be rational but may correspond to the *interests* of an individual, proposes a far more inclusive criterion for moral considerability:

Human beings who are neither self-determining nor rule-following... might have ends at which someone might reasonably be expected to aim even if he is in fact unable to choose these ends for himself, formulate purposes, plans and policies of his own, or govern his conduct by rules... we may show respect even for such human beings because it is in their interest that certain ends be attained; in that sense we may make their ends our own... it is false that 'respect' 'has been defined to fit the concept of person' and that ' "a person" is the formal object of *agape*'.

It is Watson's (1980, p. 60) conviction that the wider the range of characteristics that are deemed as having relevance and significance, the greater the number of individuals that we thereby have some moral reason to care for:

If our client possesses the capacities of a *person* we have one moral reason; if he possesses this distinctive endowment of a human being and other valued capacities, we have more than one moral reason to care. And if our other valued characteristics include some possessed by the severely mentally handicapped, we have moral reason to care for clients whom mere respecters of persons have no moral reason to help.

Watson (1980) argues that *respect for persons* provides negligible moral reason for the care of non-persons, and is an unacceptable and deficient moral principle for social work. It is Watson's (1980, p. 61) belief that the attitude of compassion towards non-persons cannot on its own secure moral considerability, for 'such compassion entails adopting a moral principle recognising them as worthy of care in virtue of their possession of certain valued characteristics *other than* those distinctive of a person'.

The principle of *respect for human beings* grounds respect for non-persons by virtue of reference to a wider range of characteristics of value; as Watson (1978, p. 41) observes, 'Respect for human beings allows, but respect for persons strictly does not allow respect for human beings who are not persons.' The latter allows for extension of indirect respect, which is qualified and derivative in nature. There is nothing inherently valuable in *this* or *that* individual, rather we condescend to extend or ascribe value, supposedly only because *we* know what it is to value ourselves and others of our kind who are of inestimable worth. Jesus (*John* 13: 34) does not direct us to love and respect only *persons*, or suggest that our love for those who fall short is to be extended as a concession – 'Believe me, when you did it to one of the least of my brethren here, you did it to me' (*Matthew* 25: 40).

The principle of *respect for human beings* is but a modern articulation of the love of humanity that inspired pioneer social workers (Woodroofe, 1971), and corresponds with the principle of *individualisation* that has been at the heart of social work from its beginnings. Discernible in the writings of Octavia Hill, Mary Richmond and Jane Addams, Robinson (1930) depicts it as 'the foundation of modern case work', being the right of all human beings to be treated 'not just as *a* human being but as *this* human being with his personal differences' (Biestek, 1973, p. 25).

A respect for human beings is borne out by Gaita's (1999) recounting of a story about a nun who cared for incurably mentally ill individuals, who in spite of their afflictions nevertheless remained, in her eyes, as *fully* our fellow human beings. The nun, in her own person, incarnates a respect for, and love of, human beings that is *transcendent* of attributes or characteristics, with Gaita (1999, p. 20) observing that 'her behaviour was striking not for the virtues it expressed, or even for the good it achieved, but for its power to reveal the full humanity of those whose affliction had made their humanity invisible. Love is the name we give to such behaviour'.

Watson (1978, p. 45) nominates 'the capacities to be emotionally secure, to give and receive love and affection, to be content and free from worry, to be healthy', with the rider that these capacities are neither definitive, nor is any particular capacity an essential condition of being owed respect. The principle of *respect for human beings* thus succeeds in grounding the moral considerability of *all* human beings, without recourse to either indirect arguments or condescension, where *respect for persons*, as traditionally conceived, so obviously and portentously fails.

The reason why we ought to care for and respect supposedly *marginal* human beings is because

> they are weak, defenceless, at our mercy. They can be hurt, injured, frustrated. We *ought* to consider their wishes and feelings, not because *we* will be hurt if we don't, but because *they* will be hurt. And the *same* goes for those creatures who are of our kind though not of our species . . . not that it offends us but that it injures sentient, appetitive, partially communicative beings.
>
> (Clark, 1978, p. 149)

Our Western tradition has a long history of identifying and isolating rationality as being constitutive of the human essence and excellence, but Rawls (1999) argues that rationality itself is a morally arbitrary property, a property that is bequeathed by nature and for which we can claim no responsibility or moral credit. In the human sphere it is commonly asserted that justice is an entitlement independent of one's capacities or excellences, and Rowlands (1998), along with Elliot (1984) and Rachels (1978), contends that our human status is also a morally arbitrary property, in the sense that it too is undeserved. We are sorely mistaken to deduce that lack of cognitive capacities or moral agency entails that we owe such individuals nothing, for to conceive moral otherness as contingent upon the possession of specific attributes, be they rational or biological, is decidedly counterintuitive (Williams, 2000); indeed we do better to view innocence as something that demands respect (Laing, 1997).

Such an expanded conceptualisation of respect cannot *a priori* exclude those numerous other animals who can, without equivocation, be said to meet these very same criteria. Accordingly, it is incorrect to insist that *respect* be cut so as to fit the conceptual cloth of human beings, or that *only* human beings are fitting objects of loving-kindness or *agape*; respect is surely due to all those individuals who have interests and ends. It must be shown that there exists a morally relevant difference if creatures with similar attributes and characteristics are to be treated in morally differential ways.

If we are to expand the range of characteristics in the manner that Watson (1978, 1980) commends, it is evident that we cannot restrict respect, so conceptualised, as being the exclusive domain of human beings without recourse to blatant anthropocentrism, for 'The catch is that any such characteristic that is possessed by *all* human beings will not be possessed by *only* human beings' (Singer, 1976, p. 265). Animals, without any fear of anthropomorphism, can be said to value many of the same capacities that we ourselves find desirable; they can be said to have interests in being emotionally secure, for contentment and absence of anxiety, to be healthy, and to give and receive love and affection. Attentiveness to the capabilities, capacities and interests of humans *and* animals is essential for their flourishing (Nussbaum, 2006).

We can also have an *active sympathy* for animals if we understand this to mean that we aim at that which *they* themselves might, for good reason, be expected to aim at. Although acknowledging that his expanded characteristics of value for human beings may also be possessed by animals, for 'They are not *distinctive* of human beings', Watson (1978, p. 47) suggests that we overcome this dilemma by reference to the principle of *respect for human beings*, for 'human beings ought to be respected for what is valuable in them'. In a practical sense, Watson's position is reflective of social work on two distinct levels. Firstly, the notion that we respect *all* human beings is a principle that can be said to inform social work practice to a far greater extent then its more abstract counterpart of *respect for persons*. Secondly, it is reflective of social work's pervasive anthropocentricity.

We shall briefly attend to the vexed moral issue of abortion, for the reason that it goes to the heart of deliberations as to which beings we might rightly term *human*. Abortion has been singled out (our treatment of the severely disabled, the senile and the insane could have likewise been addressed) for the reason that it exemplifies the ambiguity, contentiousness and equivocation that surrounds the criteria as to what it is to be *fully* human. Advocates of abortion make a clear distinction between persons and non-persons; for instance, Adams (1994) contends that foetuses inhabit a state of *nonbeing*, whilst Francione (1995b) rejects the notion that the foetus is a *discrete individual*. However the zygote I once *was* is indeed the person I now *am* (Clark, 2001), and our instinctive respect for the body of the deceased, *including* foetuses, belies the notion that it is nonsensical to respect the life of the unborn, unless we wish to maintain that respect for the former is merely out of deference to the sensibilities of the deceased's loved ones. *Pace* Adams and Francione, it is the rare parent who refers to their unborn child other than as their *baby*, specifically responding to its being and particularity.

*Any* duties conceded to foetuses are invariably conceived as indirect in nature, and we therefore ought not be overly surprised when we read that the procedure termed 'foetal reduction', a euphemism for the poisoning of a foetus by means of needle injection, is advocated as an option in the case of multiple pregnancies because it *may* enhance the health of remaining children (Stock, 2001); or justified in terms of ridding ourselves of the unborn disabled (Lawson, 1995), as a supposedly enlightened and morally uncontentious eugenics; or that abortion ought to be a matter of absolute choice (Gordon, 1977; Harrison, 1983) and morally neutral (Warren, 1973). Fears exist that the scientific discovery of growing tissue from the ovaries of aborted foetuses may well lead to a future wherein 'harvesting' of eggs from aborted foetuses and their subsequent fertilisation creates children with unborn biological mothers (Tobler, 2003), serving to validate Stephen Clark's (1977, p. 128) prophesy that 'We are already being prepared for experimentation on the aborted "products of conception", with the vague promise that so we may be saved a few more inconveniences.'

If we deem it morally permissible to abort the unborn disabled, why ought we be exhorted or be expected to respect the lives of the disabled, or why ought their wishes to live be accorded weight, once they are in the land of the living? Why ought a woman not to have the right to abort her foetus right up until the moment of birth if that is her *choice*, or why ought we to forbid a parent or parents' *right* to abortion on the grounds of say, preferred gender, intelligence, or perhaps eye colour? *We*, in our infinite wisdom, all too readily come to endorse the aborting of all those who are 'imperfect' or 'disabled', and kid ourselves that we are extolling choice, and not eugenics by stealth (Reist, 2006).

It is Francione's (1995a) belief that even *if* we assign rights to the foetus, those rights are of necessity subservient to the primary rights of the woman, and Petchesky (1984) argues that the corollary of foetuses being accorded personhood is *forced motherhood*, echoing Thomson's

(1971) equating an unwelcomed pregnancy with the woman essentially being a foetal life-support system. However, appeals to an absolute right over one's own body cannot be utilised when another body is as deeply involved as one's own (English, 1975). The notion that foetuses are not people in *any* sense is belied by the reality that they are so conceived from the moment that a pregnancy is confirmed (for instance, when a woman is referred to as 'being with child'), or that many women experience post-abortion grief (Reist, 2000), and by the fact that 'there is intense *clinical* pressure to identify the foetus as a quasi-child whose welfare the mother is obliged to foster' (Williams, 2000, p. 46). And even legally we are not unequivocal as to the moral nonbeing of the unborn, for 'Lawyers under the pressure of liberal hypocrisy may invent such legal fictions as allow a man to sue for pre-natal injuries which, if they had been more effective, would have been no wrong at all – pretending that only that is *injury* which is rationally known as such' (Stephen Clark, 1977, p. 17).

The way in which we conceptualise the moral standing of foetuses has direct implications for the sanctity of human life:

> Advocates of legal abortion cannot remain neutral about when the developing human being acquires a right to life. If they advocate abortion on request, they are implicitly valuing the claim to life of a fetus as less important than the claim of a pregnant woman to choose the size of here family, pursue an uninterrupted career, or avoid bringing up a disabled child. This may be justifiable, but it is a substantive moral position because it rejects the idea that all human life is equally sacrosanct.
>
> (Singer, 1997, pp. 85–6)

Singer (1997) argues that the moral standing of a newborn child lies somewhere between that of a foetus and that of an older child or adult:

> If there were to be legislation on this matter, it probably should deny a full legal right to life to babies only for a short period after birth – perhaps a month ... the newborn baby is on the same footing as the fetus, and hence fewer reasons exist against killing both babies and fetuses than exist against killing those who are capable of seeing themselves as distinct entities, existing over time.
>
> (Singer, 1984, pp. 125, 124)

Which begs the question, do fewer reasons also exist against *killing* the severely disabled, and the profoundly and irreversibly senile and demented, who, *we* may decide, lack the capacity to view themselves as distinct entities? And who or what gives *us* such a right to play God, or to pontificate which lives are worth preserving, and what perfection entails? Once we acquiesce to the moral priority of the strong, we are all liable to be deemed (all too readily) expendable (Brophy, 1980). Interestingly enough, even those human beings who are severely brain damaged, senile or in irreversible comas (categories commonly referred to as *human vegetables*), are nevertheless understood still to be legally and medically persons (in the metaphysical sense), but as Sapontzis (1987, p. 50) observes:

> Fetuses that do not yet look human constitute the only significant group of *Homo sapiens* widely not considered persons [in the metaphysical sense]...since those who object to calling such fetuses 'persons' also often object to calling them 'humans', these fetuses cannot provide clear counterexamples to the thesis that 'person' is just another name for human beings.

And those inclined to assign the foetus an other-than-human status, confident that it removes these beings from the moral sphere, ought to think again, for as Kerr (Preece, 2002, p. 58) observes, 'Paradoxically enough, the more animal we remember ourselves to be, the weightier the theological objections to abortion and embryo experimentation might become.' We ought to respect and value our kin because they are kin (Clark, 2000a), independent and irrespective of character or talent (and kin does not refer only to creatures of *our* kind), for what is sacred and of value is the *whole* human, not specific attributes or capacities (Weil, 1986). As my grandmother never tired of observing, it takes all types to make a world; even the *least* among us have their rightful place in the sun.

John Paul II (quoted in Clark, 2000a, p. 193) deplores

> the mentality which tends to *equate personal dignity with the capacity for verbal and explicit*, or at the least perceptible, *communication*. It is clear that on the basis of these presuppositions there is no place in the world for anyone who, like the unborn or the dying, is a weak element in the social structure, or for anyone who appears completely at the mercy of others and radically dependent on them, and can only communicate through the silent language of a profound sharing of affection.

This suggests that we ought to delight 'in the actual, present properties of creatures without ordinary human speech' (Clark, 2000a, p. 271), and belies the notion that language possession is a necessary capacity for the valuing of a creature. But if we are to value 'the silent language of a profound sharing of affection' for such human creatures, we should surely extend such consideration to animals. Those who argue that abortion represents a distinct threat to the notion of the sanctity of human life more often than not conceptualise the sanctity of human life in inverse proportion to the instrumentality of animal life; it is not the sacrosanct nature of life, but *human* life, that we are exhorted to respect. If animals at the very least match the capacities possessed by such human beings, only anthropocentrism, be that of a religious or secular variety, can serve to justify non-consideration of like interests.

Unfortunately, the reality that the capacities of supposedly *marginal* humans are often, at the least, matched by other animals often leads either to a reversion to a resolute anthropocentrism, or else to a deprecation of the value of certain forms of human life. Either human life has an absolute and exclusive value, or else only *persons* possess supreme moral considerability. The evolution story teaches us that we are all kin, embedded within the natural world, and 'as beings forming a small part of the fauna of this planet, we also exist in relation to the whole, and its fate cannot be a matter of moral indifference to us' (Midgley, 1983a, p. 91). And for the spiritually inclined, perhaps the notion that we are all creatures of God, and that our dignity (human and animal alike) resides in the fact of our common origin (a point accentuated by St. Francis of Assisi and St. John Chrysostom, among others), ought to serve to correct the notion that God is somehow made in *our* image. And indeed those of a secular persuasion, in the aftermath of the much heralded and vaunted death of God, cannot suppose that rationalism permits any absolute dichotomy between human beings and other animals. Perhaps, as Hoban's (1977, p. 158) George muses, 'It's almost as if we're put on earth to show how silly they aren't.'

The principle of *respect for human beings*, laudable as it is in guaranteeing the moral considerability of *all* human beings, is, by definition, exclusive of all other animals *irrespective* of their capacities or interests. The paradox is that whereas the principle of *respect for persons* does not *a priori* exclude beings other than humans from its scope (they must however meet the paradigmatic model of personhood, which it is assumed that all other animals fail to do), the principle of *respect for human beings* does precisely this. Whereas *respect for persons* extols the sanctity of rationality, *respect for human beings* adheres to the notion of the sanctity of *human* life, which explicitly excludes all non-human animals.

Singer (1997) argues that the Western tradition is unusual, not in its assumption that human life is sacrosanct, but in its conviction that *only* human life has sanctity – 'it uses words such as *sacred* and *sanctity* readily to describe human life, but becomes embarrassed if they are used for anything else' (Midgley, 2001, p. 184). If one conceives the self as essentially interconnected with all other selves (Chapple, 1993), then the very notion that we limit our respect and compassion to creatures of our own species is a moral and metaphysical delusion (Tobias, 1991); as Tatia (2002) observes, Jainism (along with Buddhism and Hinduism) reveres *life* rather than the *human* person. Similarly, Clark (1982, p. 46) conceptualises self as 'a being in the world', and Chapple (1993, p. 19) goes so far as to suggest that our understanding of violence and nonviolence can be intrinsically conceptualised as the interplay between self and otherness – if we conceive others *as* self, rather than the self as standing in opposition to others, 'It serves to free one from the restricted notions of self and to open one more fully to an awareness of and sensitivity toward the wants and needs of other persons, animals, and the world of the elements, all of which exist in reciprocal dependence.' The Golden Rule specifically enjoins us to see and treat all others as we ourselves would wish to be seen and treated.

The idea of human rights has come increasingly to be seen as being indispensable for talking about the world we inhabit, having what Midgley (2001, p. 164) terms an 'implacable force'. As a concept, it has served to expand our sense of responsibility to encompass all humankind, thereby immeasurably widening the scope of morality, but we nevertheless often ignore the fact that duties 'formulate the requirements towards which declarations of rights gesture' (O'Neill, 2002, p. 29). The very concept of human rights, Ife (2001, p. 15) argues, is by definition anthropocentric, but he insists that concern for the rights of animals is fully consistent with a commitment to those of human beings, and that the latter ought not to be construed as having perpetual and untrammelled right of way:

> where there is a conflict of claims of rights, the rights of the weak and vulnerable should prevail over the rights of the more powerful, and this can readily be applied to our obligations to non-human species as well. It is therefore simplistic to set up anthropocentric human rights and an ecocentric view of animal rights as necessarily in competition.

Whilst it is true that human rights are by definition concerned with human beings, it is incorrect to conceptualise such interests as anthropocentric. By definition, anthropocentrism *precludes* an equal consideration of equal interests, committing us to 'viewing and interpreting *everything* in terms of human experience and values... assuming man to be the final aim and end of the universe' (Macquarie Dictionary, 1990, p. 113; emphasis mine). Unless the rights of human beings are to be formulated and articulated in isolation from the rest of creation – and this would seem to be specifically contrary to Ife's intentions – we would be better served by the language of moral rights, which refer to different needs and interests, but which are necessarily inclusive in its scope – 'assigning a right implies the capacity protected by it should be positively nurtured, since it would be odd to protect a capacity and then be blithely indifferent as to whether or not it flourished' (Eagleton, 1998, pp. 296–7). And at its most basic level, Regan (2004, p. 9) argues that the notion of animal rights 'means only that animals have a right to be treated with respect'.

Ife's (2001) differentiation between *anthropocentric* human rights and *ecocentric* animal rights evokes Nozick's (1995, p. 39) observation of 'utilitarianism for animals, Kantianism for people'. There exists, Brophy (1979, p. 72) avers, 'a necessary continuity between the rights of all animals, as animals, including human animals in with the others', and

> It is of little use to claim 'rights' for animals in a vague and general way, if with the same breath we explicitly show our determination to subordinate those rights to anything and everything that can be construed into human 'want'; nor will it be possible to obtain full justice for the lower races so long as we continue to regard them as beings of a wholly different order, and to ignore the significance of their numberless points of kinship with mankind.
>
> (Salt, 1980, p. 9)

We have come to realise, albeit gradually, that we have duties and obligations that transcend consideration of the interests of our own species, but because the language of rights has traditionally been seen as the sole preserve of humanity, attempts to utilise 'rights' and 'animals' in the same breath have met with both incredulity and resistance, despite the fact that the emphasis accorded to human claims was not originally intended to preclude concern for other terrestrial creatures. Indeed,

> 'the idea of dismissing them as mere disposable instruments now strikes many of us as immoral and repulsive. Yet the rationalist half of the tradition is deeply committed to claiming that only rational

beings of a strictly human kind can have the kind of value or importance that would bring them within the scope of morality at all'.
(Midgley, 2001, p. 161)

We earlier noted that Ife (2001) maintains that a human rights perspective necessitates that where conflict of claims of rights exist, then
the rights of the weak and vulnerable should prevail, and furthermore he asserts that this principle can be readily extrapolated to
animals, and that human rights need not take precedence over the
rights of animals. Given Ife's conviction that the weak and vulnerable have a moral priority, it can be argued that the claims of animals
ought thereby to be accorded precedence when they conflict with the
rights of human beings. This notion that we might owe animals *more*
than equal consideration derives from the fact that 'the weak and the
oppressed constitute a special moral category of moral obligation based
on our special relationship of power over them' (Linzey, 1994, p. 36).
Indeed, Linzey (1994, pp. 41–2) contends that the historic language
of justice, and related ideas of equality and rights, indispensable as
they are,

> fall short of meeting the moral claim that animals make on us... It is
> quite possible for a parent to respect all the rights of a child and
> still be less than what a parent should be. He or she may not beat
> the child, provide as far as possible for its education, even care for
> it in times of sickness and need, and yet may still fall short of being
> the loving, forgiving parent that all children actually need for their
> growth into adult personhood. We should certainly speak of the
> rights of children – as we should speak of the rights of animals – but
> what parents owe children – as human[s] owe animals – goes beyond
> the respecting of their rights.

Rather than human excellences or supposed uniquely possessed capacities, or human species membership *per se*, having the definitive say in
moral deliberations, it is the individual (human or animal) that takes
centre stage; the greater our capacities the greater our obligation to all
that lives (not the greater *our* rights), the greater the weakness and vulnerability of those human and non-human animals the greater *their*
claims to moral priority.

Social work has a long and admirable history of continually reiterating
that we have a *special* duty to the weak and the vulnerable, a duty that
is morally obligatory, neither supererogatory nor optional. As noted in

the opening chapter, the solicitude of nineteenth-century social reformers in Britain and America for the vulnerable transcended the species barrier, and amongst some early social workers concern for non-human animals was seen as perfectly compatible with regard for members of our own species. Remarkably, this ready recognition of the universality of compassion and loving-kindness was at its zenith. Contemporary social work remains resolutely wedded to an anthropocentric worldview, the reality of our biological kinship, and the fact that we share capacities in common with other animals, seen as having no moral implications. This dearth of moral imagination results in the turning of many a blind eye to the welfare and well-being of animals encountered in practice; as the contemporary social worker Lynn Loar (1999, p. 120) observes, 'Sensitive, thoughtful, and caring people seem rarely, if ever, to have considered the problem of cruelty to animals and its ramifications for human interaction.'

And yet slowly but surely awareness of the interrelationship between human and animal welfare is coming full circle, and again beginning to be understood as mutually inclusive concerns. It has been shown that there is a strong correlation between children who have shared their lives with companion animals in childhood and later adult concern for the welfare and well-being of animals (domestic and non-domestic) *and* human beings (Paul and Serpell, 1993); that there exists an interrelationship between feelings of empathy for humans and animals (Paul, 2000), and an unambiguous linkage between animal abuse and later aggression against human beings, particularly violence against women and children (Ascione and Arkow, 1999; Linzey, 2009a; Lockwood and Ascione, 1998). Notwithstanding Hutton's (1983) suggestion that animal abuse might well be efficacious as a diagnostic approach in social work, there nevertheless remains resistance to the linkage between animal abuse and domestic violence (Grant, 1999).

The importance of animal companionship is likewise acknowledged (Anderson et al., 1984; Beck and Katcher, 1996; Podberscek et al., 2000), especially for the elderly (Lago et al., 1985), with specific benefits for human health – particularly cardiovascular (Friedmann et al., 2000) – and human mental well-being (Bustad, 1980). Animals also play a salient role in the emotional and moral development of children (George, 1999), and in work done with children from violent and/or sexually abusive backgrounds (Roseberry and Rovin, 1999).

Membership of the moral community cannot be restrictively narrowed to being the exclusive preserve of rational adult human beings, for

What rights there are, in short, are to be found by discovering what all creatures might, without self-contradiction, be required to do or refrain from doing... If we abandon the arbitrary line that has been drawn around the human species and understand that all creatures are members of the one community, the terrestrial biosphere, we can ask what rules might obtain in that system.

(Clark, 1985c, p. 21)

If we desire to persevere with the language of personhood, then perhaps we ought to conceive it in much the same manner as Darwin (1936) conceives evolution, a continuum marked by difference in degree. Personhood, Smuts (1999, p. 118) insists, is not contingent upon the attribution of human characteristics to other animals, for intersubjectivity transcends the species boundary:

they are social subjects, *like us*, whose idiosyncratic, subjective experience of us plays the same role in their relations with us that our subjective experience of them plays in our relations with them. If they relate to us as individuals, and we relate to them as individuals, it is possible for us to have a *personal* relationship. If either party fails to take into account the other's social subjectivity, such a relationship is precluded... personhood connotes a way of being *in relation to others*, and thus no one other than the subject can give it or take it away. In other words, when a human being relates to an individual nonhuman being as an anonymous object, rather than as a being with its own subjectivity, it is the human, and not the other animal, who relinquishes personhood.

Ultimately, our own moral well-being is dependent upon our capacity to acknowledge selfhood in others (Lowry, 1999), for persons are not Cartesian selves, but exist in relationship with others; Griffin (1984, p. 165) observes that

Learning to recognize individual animals under natural conditions, a very important advance in ethology, has led to the discovery of previously unsuspected patterns of behavior that could not be appreciated when the animals were treated as interchangeable units. One discovery has been that not only the ethologist but the animals themselves often recognize their companions as individuals and treat them accordingly.

What is needed is a more holistic model of moral rights that transcends an exclusive, or primary, focus upon humankind. Moral rights ought to be extended to all beings that have interests, and this entails a respect for individuals *independent* of their capacities or species membership. Regan (1983, p. 327) avers that the 'principal basic moral right possessed by all moral agents and patients is the right to respectful treatment... [all are] viewed as having a distinctive kind of value (inherent value) and as having this value equally'. We can therefore jettison the principles of *respect for persons*, and *respect for human beings*, in favour of the more morally inclusive and biologically sound *respect for individuals*. As Clark (1997, p. 104) observes, it is the individual dignity of each subject-of-a-life, not only rationality, that is deserving of our respect, and 'we treat such creatures wrongly when we systematically deny to them the sort of respect and care that we are so eager to have lavished on ourselves'. We assuredly wrong other creatures when we deny that their lives are *theirs*, when we cause them pain and distress, deny them the lives they would otherwise lead, and treat them as though they were nothing more than our *things* (Clark, 1997). Moral patients, humans *and* animals alike, are owed a right of recipience, and a right to respectful treatment is universalisable for both moral agents and moral patients, for 'any principle, or prince, that accords rights to the weak of our own species must also accord them to animals' (Stephen Clark, 1977, p. 34). It is for such reasons that Regan (2004, p. 38) avers that 'Moral rights breathe equality.'

In making the observation that a sharp dichotomy between means and ends is deeply embedded in the Western ethical, political and psychological vocabulary, Iyer (1973, p. 361) comments that 'the dangerous dogma that the end entirely justifies the means is merely an extreme version of the commonly uncriticized belief that moral considerations cannot apply to the means except in relation to ends, or that the latter have a moral priority'. It is Gandhi's (Iyer, 1973, p. 361) conviction that there exists an inviolable continuity between means and ends – 'The means may be likened to a seed, the end to a tree; and there is just the same inviolable connection between the means and the end as there is between the seed and the tree.'

The implications of the view that ends have primacy over means is borne out by Singer's (1984) observation that a utilitarian *must* endorse the position that possibly painful experiments conducted upon those animals, who far as we can ascertain are neither rational nor self-conscious beings might, under certain conditions, be justified were a majority of others to be benefited as a consequence – 'it is at least arguable that no wrong is done if the animal killed will, as a result of the killing, be replaced by another animal living an equally pleasant life. Taking this view involves holding that a wrong done to an existing being can be made up for by a benefit conferred on an as yet non-existent being' (Singer, 1984, p. 104). It is *always* possible, Linzey (1987) argues, to justify the utilisation of animals for experimental purposes on the grounds of utility, if for no more substantive reason than the fact that *nothing* can be conclusively proved to be useless.

Singer's conclusion reflects the inherent conflict that is a consequence of the fundamental incompatibility of the principles of equal consideration and the principle of aggregation of interests (Rowlands, 1998). The utilitarian calculus conflicts with the rights perspective, which, whilst not immune to consideration of consequences, nevertheless insists that they cannot be the *only* basis on which moral decisions are to be reached. The rights view rejects the notion that individuals can be considered as receptacles of value, and appeals to aggregate considerations ride roughshod over the notion that each individual has value in and of themselves, for it is their uniqueness that grounds their irreplaceability (Cave, 1982).

Where conflicts occur, Regan (1983, pp. 305, 308) proposes the *miniride* and *worse-off* principles as means of arbitration; the former decrees that

Special considerations aside, when we must choose between overriding the rights of the many who are innocent or the rights of a few who are innocent, and when each affected individual will be harmed in a prima facie comparable way, then we ought choose to override the rights of the few in preference to overriding the rights of the many.

Whilst the latter entails that

Special considerations aside, when we must choose to override the rights of the many or the few who are innocent, and when the harm faced by the few would make them worse-off than any of the many would be if any other option were chosen, then we ought to override the rights of the many.

The principle of *respect for individuals* insists that each individual subject ought to always be treated as an end, and this, in Weber's (1991) opinion, is the essence of Buber's (1970) distinction between an *I–It* and an *I–Thou* relationship. The violence inherent in any *I–Thou* relationship can be transformed via respect, admiration or affection, 'or one of the countless attitudes that men call love' (Buber, 1970, p. 17). For Gandhi *ahimsa* is the means, whilst Truth is the end (Chatterjee, 1985), and Bondurant (1965) draws our attention to the conceptual parallels between *ahimsa*, Christian charity and *agape*. Indeed Gandhi characterises *ahimsa* as *love* (Jesudasan, 1984), a love not merely of the human community, but of the community of all living creatures (Altman, 1980; Lodrick, 1981) as well as the natural world (Chapple, 2002).

In conclusion, we have already seen that there exists no morally relevant characteristic that is possessed by *all*, and *only*, human beings. The principles of *respect for persons* and *respect for human beings* are evidently inadequate in grounding our moral obligations to other subjects-of-a-life, for what matters is not whether a being is a *person*, nor whether a being is a *human* being, but rather whether one is a creature who is sentient, has interests, is an experiential subject with an individual welfare, and for whom life can fare ill or well for them over a period of time (Regan, 1983). Respect ought to be extended to all sentient creatures, human or animal, whatever characteristics or capacities they possess, and all ought to be included in the kingdom of ends, in order that 'pain to all upon it, tongued or dumb, shall be kept down to a minimum by loving-kindness' (Hardy, 1976, p. 558). To do so recognises our ontological and moral continuity with all other species, what Moore (1992) refers to as the *universal kinship*, so that we may, in time, come to approximate Gandhi's (Wynne-Tyson, 1985, p. 92) ideal that 'I want to recognize brotherhood or identity not merely with the beings called human, but I want to realize identity with all life, even with such things as crawl upon earth.'

*Respect for individuals*, which enjoins and entails respect for, and loving attention to, all sentient creatures, is a principle that ought to inform social work's moral framework if it is to have any pretension to being a holistic discipline. If the central concept of morality is *not* respect for rational, adult human beings, but, as Murdoch (1996, pp. 30, 38) suggests, 'the individual knowable by love...which connects morality to individuals, human individuals or individual realities of other kinds', then the principle of *respect for individuals* enables such an extension of moral considerability.

As the Qur'an (Wynne-Tyson, 1985, p. 139) reminds us,

There is not an animal on the earth, nor a flying creature on two wings, but they are people like unto you.

# 5
# A Morally Inclusive Social Work

Few people seem to perceive fully as yet that the most far-reaching consequence of the establishment of the common origin of all species, is ethical; that it logically involved a readjustment of altruistic morals by enlarging as a *necessity of rightness* the application of what has been called 'The Golden Rule' beyond the area of mere mankind to the whole animal kingdom... The discovery of the law of evolution, which revealed that all organic creatures are of one family, shifted the centre of altruism to the whole conscious world collectively.

*Thomas Hardy* (1930, pp. 141, 138)

A social work code of ethics worthy of the name ought to stress, Wilkes (1985, p. 45) avers, 'the law of general beneficence – of never doing to others what you would not like them to do to you', and *others* cannot be only those of *our* kind. The substantively revised *AASW Code of Ethics* (1999b) in the addendum following this chapter provides social work practitioners with the moral framework and conceptual wherewithal to respect and attend to all sentient individuals, irrespective of their species membership. By way of contrast to the existing Code (and by implication, *all* contemporaneous codes), this change in emphasis has profound consequences for social work thinking and practice. It expands social work's moral and ethical sphere, but in no way detracts from nor enervates social work's core mission, nor represents an affront to human dignity. Rather, it acknowledges our terrestriality and our ontological continuity and kinship, and the moral and ethical implications that issue from such recognition.

It also grounds *respect for the individual* as the bedrock social work principle, and enjoins us to respect human beings *and* our fellow animals as members of the moral community. This chapter will tease out some of the practical implications flowing from the adoption of this principle, via reflection upon the case scenarios outlined in Chapter 1.

It may well be that our treating animals *like* animals in the traditional instrumental sense predisposes us, as Aquinas (1989) and Kant (1990) warn, to be cruel to our fellow humans. If it could be conclusively proved that cruelty to animals *never* precipitated like treatment of my human fellows, I would then be at complete liberty to make the lives of my animals a living hell, should I so desire (since I would therein not be offending against anyone else's property rights), and still be considered a perfect gentleman in all manner of human affairs. If we act towards sentient creatures in ways that either cause avoidable suffering or ride roughshod over their interests, then there is good reason to believe, indeed fear, that we will become desensitised in our dealings with other sentient beings – 'we may take it for granted that, in the long run, as we treat our fellow beings "the animals," so shall we treat our fellow-men' (Salt, 1935, p. 51).

Locke (1990, p. 119) argues that one's attitude and behaviour, compassionate or otherwise, towards humans and animals, has its foundation in childhood, and whilst cruelty is not natural, 'the custom of tormenting and killing of beasts will, by degrees, harden their minds even towards men; and they who delight in the suffering and destruction of inferior creatures, will not be apt to be very compassionate or benign to those of their own kind'. It is indeed remarkable that we should ever have thought that we might do as we please with our animal kin, and to suppose that it will have no direct bearing upon either our moral character or demeanour towards our fellow human beings. Indeed were acknowledgement of a difference in degree, not kind, to have no moral bearing upon our treatment of animals as subjects, we would quite rightly feel trepidation as to the treatment of human beings. But it is just as odd to suppose that the way in which we treat all other animals is nothing more than an exercise (or trial run) in human moral development, or a matter of aesthetics, and that our own fate is independent of that of all other creatures – as Albert Schweitzer (Wynne-Tyson, 1985, p. 316) observes, 'Until he extends the circle of his compassion to all living things, man will not himself find peace.'

Interestingly, there has been no parallel admonition that treating machines or non-sentient nature as things will increase the likelihood of humans being so treated. The analogy and metaphor does not hold precisely because animals are like us in ways that machines and non-sentient nature are not, that sane and humane people consider it evil to inflict unnecessary pain or harm, or to occasion neglect. For as Clark (1997, pp. 158–9) observes, 'it has never been easy to explain quite why such cruelty or negligence is wrong...If humans matter so much more than non-humans it must be that they are of so radically different a kind that it is hard to see that "hurting animals" is of the same kind as "hurting humans".'

The notion that whilst we ought to be kind to animals, we owe them only indirect duties, leads Salt (1980, p. 133) to remark that 'kindness is, so to speak, the water, and the duty is the tap; and the convenience of this arrangement is that the man can shut off the kindness whenever it suits him to do so'. Too many of us have extended families and wider households that include myriad animals for such arguments to hold water completely. If our much loved dog Tess, for instance, is set upon by a human neighbour, we would be considered rather odd, to say the least, if we were to view it solely as an infringement and abuse of our property rights, or to interpret the action either as an indirect attack on ourselves or as enhancing the prospect of the actual assault of fellow humans, perhaps, a little understandably, ourselves. We would primarily be affronted because of the harm and hurt visited upon Tess. Even were it categorically ascertained that this action would not lead to fellow humans being so treated, or even if it were not *our* dog but *yours* that was harmed, we ought still to feel aggrieved for the animal's sake. We would undoubtedly sympathise and empathise with you, but once more this would be the consequence of a shared outrage at what happened to the animal first and foremost.

Earlier chapters canvassed the traditional claims that have been put forward to warrant either the absolute or relative dismissal of the moral claims and standing of other animals (consciousness, language, moral agency, rationality, sentience, soul and species membership), and found them to be variously arbitrary, erroneous or irrelevant in excluding other animals from the circle of moral considerability. These arguments have been seen to be incommensurate with, and an inadequate basis upon which to construct, the edifice that is social work's ethical and moral framework. We have now arrived at a position whereby a substantial case has been mounted for the adoption of the species-*inclusive* code of ethics, which grounds and secures the moral considerability, and respects the interests and well-being, of all sentient creatures. The absence of *any* discussion of the moral nature of human dealings with animals leaves social workers manifestly ill-equipped to respond to animal neglect or abuse, let alone to accord due consideration to the interests of other animals; Loar (1999, p. 120) characterises this pervasive attitude within social work as 'I'll only help you *if* you have two legs' [emphasis mine].

There is absolutely nothing in the curriculum or education of social workers that prepares them to be able to resolve the conflicts, or to negotiate their way through the moral dilemmas, posed at the commencement of this book, for there is no linkage made between the maltreatment of human beings and other animals. Indeed the very suggestion is seen as intrinsically demeaning to human beings, and concern for animals in the midst of human travails is often seen as evidence of misplaced priorities (or worse, an indulgence) and misuse of limited time, energy and resources. This in spite of the fact that studies consistently show a strong correlative relationship between the abuse of human beings and animals.

Social workers (apologies to those who beg to differ) do not conceive human beings and other animals as kindred creatures. And it is *this* metaphysic that makes *all* the difference in our response to practice situations and dilemmas of the kind related at the commencement of Chapter 1. We may feel unease, we may feel alarmed, we may even feel guilt, but so long as we conceptualise other animals and their interests as disparate, peripheral and invariably subsidiary to those of their human counterparts, we shall remain in the thrall of a moral paralysis. So long as we remain blinkered about the scope of moral relationships, and the reality of moral identities that exist between individuals (Watson, 1980), human *and* non-human, that morality entails an awareness of other individuals, and that we act morally *when* we take account of the interests of individuals (Todorov, 2000), we shall continue to be beholden and wedded to an instrumental and anthropocentric worldview, and will fail to attend to them with the respect they are due.

What is often missed in the linkage between violence to animals and violence to human beings is the fact that a fellow sentient creature has been *directly* harmed; violence towards or neglect of animals is conceptualised *indirectly* (befitting their status as *things*), whilst violence towards or neglect of, human beings, is conceived *directly* (befitting their status as *persons*). A case study highlights this phenomenon:

> In 1992, 12-year-old Eric Smith killed a neighbor's cat. He was made to apologize and do some yard work for the wronged neighbor. In 1993, Eric Smith killed a four-year-old boy. He was convicted of second-degree murder for that offence the following year. Shortly after Eric Smith's trial, I attended a workshop given by the child psychiatrist affiliated with Yale University who testified on the child's

behalf at his trial. I was impressed by the psychiatrist's compassionate yet objective assessment of the boy. At the end of the workshop he took questions, and I asked if anyone had bothered to report the killing of the cat to the local humane society, or municipal animal control agency. He replied that although that had not been done, consequences had been imposed on the child, namely the apology and yard work. I responded that the consequences addressed the property damage the neighbor experienced in the loss of his cat, but not the boy's taking the life of a sentient creature. Indeed, had the crime been reported, the child would likely have been required to undergo counseling and to have supervision when around defenseless living creatures. The psychiatrist replied that nothing more than the restitution and apologizing were done and he agreed that something important had been missed. The response of this prominent and competent psychiatrist demonstrates the lack of awareness of the significance of cruelty to animals common among both human service professionals and the general public.

(Loar, 1999, p. 120)

In such a scenario, it would be all too easy to *completely* overlook the killing of the cat in light of the fact that a child was subsequently murdered. Indeed the child's death could be seen to be trivialised or demeaned by even raising concerns about the cat's death. As Loar relates, the infliction of injury or harm to animals, including death, is consistently seen as chattel damage, rather than as harm done to specific individual creatures. This tendency to view animals as *things*, and disposable at that, is borne out by the fact that in Australia, during the 2008–09 financial year, the RSPCA (2008–09) received 156,621 animals, of which just over 46 per cent were euthanised.

Because social work invariably views the notion of respect as relating exclusively to human beings, it has immense difficulties in grappling with how it is that we are able to speak of our responsibilities for, and duties and obligations towards other creatures. Our social workers in the case studies are at a downright loss to know *what* or *how* to think, or *how* to act, with the consequence that they invariably sit on their collective hands and do neither. When Banks (1995) asserts that the very nature of moral judgements is human centred, she is effectively speaking for social work as a discipline. It is little wonder then that social workers lack the basic conceptual understanding and wherewithal that might enable them to respond to the interests of an individual of another species, or that they interpret this neglect or absence of attention as confirmation that animals do not matter – to adapt Coetzee's (2000, p. 73) Lucy's observation, on social work's scale of priorities 'animals come nowhere'.

This is belied by the reality that human beings and domestic animals form a genuine community and that a concern for animals is an inherent feature of human moral sensibility, as is sensitivity to their suffering. In the fourth scenario, the elderly woman indisputably regards her 15-year-old dog *as* family, indeed her only immediate family member. Whilst the social worker involved is only too well aware that the woman's anxiety at her imminent transfer to the nursing home has been immeasurably compounded by the dilemma of what will happen to her elderly dog, the focus is exclusively on the impact this has upon the elderly woman and *her* welfare and well-being. In contrast, the woman, whilst not oblivious to this fact, is primarily focused upon the welfare and well-being of her beloved dog, in particular on the distress that the pending separation will cause the dog. The woman conceives the dilemma as not hers alone, and sees the proposed resolutions (either rehousing or euthanasia) as no solutions at all. She rightly sees her and her dog's well-being and flourishing as being essentially interdependent. Precisely because both the woman and her dog are exceedingly vulnerable, both are owed moral priority.

When she declares that she would rather die herself than agree to either solution, her supposed melodramatic response might be thought best addressed by gently persuading her to adjust to 'reality', with a likely assumption being that time and a surfeit of human company in her new surrounding will ultimately more than compensate for her current distress. Either of the proposed remedies are impositions, and assume that others know what is in the best interests of the woman concerned. Were the social worker (and indeed members of all other disciplines) to seek to secure the optimum outcome for the woman and the dog, this would surely entail acknowledging the emotional fellowship and interdependence that holds between them, and as a consequence forcefully advocate that nursing home facilities should cater for residents and their animals as a matter of course. Social workers would have an integral role to play in drawing attention to the importance and centrality of animals in the lives of the elderly, and to advocate on behalf of them and their animals when and where necessary.

Likewise, the woman in the fifth scenario ought not to be expected to forsake her animals when she finally decides to leave her violent husband. When the social worker sympathises with her dilemma, but suggests no practical resolution, the woman's determination to otherwise remain in her unsatisfactory situation is understandable – the more so for her plausible fear that her husband will make good his threats to either harm or kill her animals if she should leave him. The social worker is understandably apprehensive should the woman remain with her husband, and conceives the issue as being exclusively concerned with the woman's welfare, well-being and safety. In contrast the woman is if anything more focused upon that of her animals (given that she has no concerns about her husband's treatment of and capacity to care for their children), and it is their vulnerability and dependency that occupies her and her decision making. Should the woman leave without them, it is no solution at all merely to report the concerns to police or animal welfare agencies, for even if the animals are subsequently removed they would in all likelihood be either impounded by the local council or placed in an animal refuge. Either solution would only serve to further distress the woman concerned, with the best outcomes being that the animals are either claimed by the woman if and when she is permanently rehoused, or that the animals themselves are given new homes or else possibly euthanised. Were the social worker to attend to the vulnerability, welfare, well-being and safety of both the woman and her animals, he or she would surely be forcefully advocating that services responding to women in these and similar situations must make a holistic response and address the needs of both human and non-human individuals, for *both* their sakes. At the least this would entail advocating for the short-term funding to assist in the boarding of animals until such time as they were able to assist the women to find permanent accommodation. Whilst it is acknowledged that an increasing number of agencies in Australia have secured government funding to assist victims of domestic violence with their animals (or that on occasion the animals of children placed in protective care have been temporarily accommodated), the responses nevertheless remains more piecemeal or *ad hoc* in nature rather than an integral aspect of a holistic approach. Furthermore, the majority of such responses remain exclusively focused upon the psychosocial well-being of the humans involved.

Social workers are regularly confronted with similar dilemmas, and they, better than most, are intimately aware of the emotional fellowship and interdependence that exists between men and women and their animals. Similarly, children placed in care have their emotional distress immeasurably compounded by their separation from cherished animals. This knowledge needs not only to be acted upon in crises, but utilised in the education of practising social workers and in undergraduate social work courses. Needless to say social workers will of necessity have to develop and cultivate linkages and relationships with animal welfare agencies, veterinarians and ethologists in order to ensure the best out-comes for the animals concerned. A significant aspect of this process will be the acquiring of knowledge from these and other disciplines so as to enable social workers to better understand the underlying structure of needs and dispositions of the animals they come across in practice; this will allow them to utilise this knowledge so as to better assist and work with individuals and families in the resolution of dilemmas regularly encountered in social work practice.

Social work has, from its genesis, been imbued with a moral passion to accord moral priority to the weak and the vulnerable, and has sought to champion and defend the dignity and worth of each and every human being, but in so doing it has assumed that *only* human beings matter. However, individuals like Maria Dickin saw nothing inherently con-tradictory or demeaning in making linkage between our treatment of human beings and our fellow animals, and one can but speculate as to what Dickin might have had to say to us about contemporary social work's bifurcation of human beings and animals into morally disparate categories. In common with celebrated humanitarian reformers, both her predecessors and contemporaries, Dickin would surely be dismayed at present-day social work's moral myopia in relation to other animals. Faced with the privation and penury that constituted her work with the working class poor, Dickin, by contemporary lights, could well have been excused (indeed commended) were she to have concerned her-self *exclusively* with their concerns and welfare, and channelled all her energies and finite resources into seeking to alleviate their conditions. An individual like Maria Dickin could be seen as little more than a Victorian relic, at best an eccentric, but more likely than not as represen-tative of those *animal lovers* who place higher value upon non-human animals, and who are either lacking or unappreciative of human com-panionship (Clark, 1996b). The notion that Dickin's concern for the animals of the poor could be seen as a case of misplaced priorities is itself, at worst, the consequence of an exclusive humanism or, at best, the product of thinking that compassion is a finite and irreplaceable resource that ought be rationed to more deserving and worthy cases. Consequently social work exhibits a routine tendency to view the inter-ests of human beings and those of other species as being essentially competitive:

> The problem of competition presents itself to many people in a form more or less like this: Must we really acknowledge all our long-lost cousins and heave them into the humanitarian lifeboat, which is

already foundering under the human race? Or can we take another look at the rule-book and declare the relationship too distant, so that we are justified in letting the whole lot sink?

(Midgley, 1983a, p. 19)

A lifeboat model of morality makes it difficult for us to make sense of our responsibilities and duties, not only to other animals and the natural world, but to distant or indigent human beings (Hardin, 1979; Singer, 1972). If one conceives compassion and moral considerability as being exclusively a matter of competition, then it becomes all the easier to refuse to grant consideration to the interests of those who are not near and dear to us. When it comes to the good ship Earth, social work employs a lifeboat model that divvies up humankind and all other species into disparate moral realms; whilst averring that all human beings are in the same boat, humans are conceptualised not only as first class voyagers, but as the *sole* passengers on the moral high sea.

*Being kind* to animals, or more likely refraining from cruelty, are injunctions unlikely to elicit or broker any meaningful objection or opposition, but the notions of kindness and cruelty are themselves problematic; both have

conceptual connections with 'the mind of the agent' – namely, with the agent's motives and intentions ... [But] the morality of what persons do (the rightness or wrongness of their actions) is logically distinct from, and should not be confused with, their 'mental states', including the motives or intentions from which their acts proceed.

(Regan, 1982, p. 69)

The moral rightness or otherwise of an individual's actions is separate from how that individual *feels* about the suffering of an animal or a human being; for example, we rightly condemn cruelty to children, but do not limit our condemnation *only* to those cases where it can be ascertained, beyond all reasonable doubt, that the perpetrator obtains enjoyment or pleasure from either causing pain or harming children. It is, however, unfortunate that our relations with animals are more often than not framed in terms of cruelty; sadists and psychopaths aside, the overwhelming majority of people who cause suffering in animals do not do so because they thereby obtain enjoyment or gratification. Rather than deeming motives to be decisive, Rowlands (2008) argues that evil invariably entails a failure of our moral (protecting the defenceless) and epistemic duties (subjecting our beliefs to apposite critical scrutiny).

To return to our first casework example, the beleaguered mother of four does not keep the family's two dogs restrained in such substandard conditions because by so doing it provides her with one of the few remaining pleasures in her otherwise impoverished, joyless and wearisome existence. Even so, the very fact that the woman is not intentionally cruel does not in itself relegate the treatment of the dogs to moral insignificance; the need for moral judgement is not less necessary because of the absence of any ill intent. The point of the matter is that we *owe* it to animals to treat them in ways that respect their inherent value, individuality and subjectivity, as something *due* to them. Likewise, in the second and third casework scenarios the neglect and deaths of animals again might not have been the result of malevolent intent, but the consequences are none the worse for this being the case.

By way of further explicating this observation, we might consider the following – if we were to examine a social worker's response to children who were neglected, but not through any premeditated or ostensible intent to cruelty on behalf of their parent/parents, we would nevertheless trust that the social worker would quite rightly feel that the neglect was still an issue that had to be attended to, irrespective of the absence of any malevolent ill-intent. We would *expect* this to be the case, not merely because of statutory requirements (after all, we don't always need to be *told* to do what is right), but because it is morally incumbent upon us to act in the interests of the children's welfare and well-being; as Garner (2002) argues, moral convictions are more critical than legal compulsion. Neglect clearly violates the individual's right to be so treated, and it is to this that the social worker attends; we *owe* it to children to treat them in ways that are respectful of their inherent value. Accordingly the social worker would conceive that a significant part of their role is to provide ongoing support to the family and practical parenting skills in order to maximise the welfare, well-being and flourishing of *all* family members.

Regan (1991) argues that Kantianism, utilitarianism and contractarianism all fail to show why such abusive or neglectful behaviour or treatment is morally wrong, as do appeals to species membership. Whilst Kant's (1964) categorical imperative would appear to ensure that we always act in ways that are universalisable (I cannot acquiesce to the neglectful treatment of all other children and expect that an exception be made in the case of my own children), Regan observes that *if* I am prepared to assent to my own children being treated as mere means, and every other parent in the world accedes likewise, the universal law would apparently see this as permissible because it is universalisable.

The utilitarian framework, in all its variations and guises, seeks an aggregation of interests (exactly how this is to be achieved is rather unclear); by this calculus, the pleasures and enjoyment obtained by neglectful parents might outweigh the suffering and harm experienced by their children. Utilitarianism cannot tell us why the neglect is morally wrong, given that it sees no difference between acts of *commission* and *omission* (Clark, 1997).

Contractarianism discriminates against all who are not rational, adult contracting agents. Precisely because children (the more so the younger they be) are not in a position to be aware of what is in their best interests or possess the capacity to enter into contracts, their welfare and protection is dependent upon what contracting adults deem is in *their* self-interest – if neglect serves the self-interest of contracting adults then it receives the seal of contractual approval (Regan, 1991).

Finally, Regan contends that the argument that children are human beings fails to enlighten us as to why treating children in a neglectful manner is wrong, given that species membership has neither moral relevance nor decisiveness.

Where a social worker *does* attend to the neglect or abuse of animals, it will more than likely be suggested that any concerns be reported to either the police or an animal welfare agency, and in so doing has fulfilled his or her duty. Apart from the fact that it is highly unlikely that any social work agency has a specific policy or guidelines as to the appropriate response to such situations, this suggestion confirms that beyond a minimal response, social workers have no further obligations. Even if it comes to pass that social work agencies were to make the reporting of animal abuse and neglect mandatory, that would be but half of the story. It is not enough that animals are removed, or that individuals are prosecuted for their abuse and/or neglect, given that animals (if not those whose abuse or neglect has been reported) will in all likelihood always be part of the households of those families; any more than it is enough that children are removed from abusive or neglectful parents, or that a violent spouse or partner is removed from the family home. Social workers have a far wider brief than a merely mandatory or crisis management role – except in genuinely exceptional cases, the ideal is either the retention within or return of children to their families, and with couples the acquiring of interpersonal skills that assist in the non-violent resolution of conflict. In both cases, social workers are uniquely placed to work with individuals and families, ideally assisting in the recognition that the wrong done is essentially a failure to respect the preciousness and uniqueness of individuals wronged.

Likewise, social workers have a critical role to play in both supporting and educating individuals and families when confronted with the moral dilemmas like those in the casework scenarios, one congruent with their more traditional role of working with human individuals. Whereas other intervening agencies and disciplines are focused more or less exclusively upon the welfare and well-being of the animals involved and the prosecution of those humans held responsible for their abuse or neglect, a response informed by legal as much as moral requirements, social workers are uniquely placed to respond in a more holistic manner. Whilst responding to the initial instances of abuse and neglect is critical, it is also imperative to work with those human individuals and families in order to effect moral as well as practical welfare changes, for the good and well-being of both human and non-human animals, both in the present and the future. And in such an important task social workers, in making informed moral judgements (which, by their nature, are species *inclusive*), are less likely to be judgemental in their dealings with the humans involved, and consequently more likely to effect an awareness of the dependency and preciousness of their animal family members.

Codes of ethics quite rightly insist upon the highest ethical standards in our dealings with our fellow humans, but this much done, social workers rest upon their laurels, content in the belief that this exhausts their moral obligations. That moral judgements are deemed to be exclusively concerned with human subjects serves only to highlight the moral inconsistency and compartmentalisation at the heart of contemporary social work. In their private lives social workers respond to *their* dogs or cats as conscious, sentient and subjective creatures with their own unmistakable individuality and preciousness, and would never countenance anyone maltreating, killing or eating them, or acquiesce in their use as research tools. Nevertheless in their public roles their moral attitudes towards nonhuman animals *per se* remains steadfastly Laodicean, for the love and solicitude accorded to their own fails to translate into any broader moral concern or normative implications.

The very fact that untold numbers of animals are routinely sacrificed worldwide in the pursuit of human health and well-being rates nary a mention in social work literature, and no social work code of ethics, or social work ethics groups, even acknowledge, let alone debate the morality of the issue. Social workers have worked in health and medical settings for almost the entire duration of the discipline's existence – in the Australian context, McCormack (2001) reports that health and community services employ approximately three-quarters of all social workers – and yet one will search in vain for *any* mention of the moral propriety or otherwise of the routine utilisation and exploitation of animals in all manner of medical research. It is estimated that 100 million vertebrates are used annually worldwide (BUAV, 2010), with 7.1 million animals used for research and teaching purposes in Australia (Humane Research Australia, 2010), it being routinely assumed that the potential benefits for human beings ensuing from such research invariably override any moral objections or qualms. But not all knowledge or benefits are worth the price that they exact from our moral sensibilities:

> what justifies the totally disproportionate cost of our presence? Ask it for once without presupposing the answer of the egotism of our species, as God might ask it about his creatures: Why should a dog or a guinea pig die an agonizing death in a laboratory experiment so that some human need not suffer just such a fate?...Why, in the perspective of eternity, should the life of a human be more precious than that of a dog? Why should the dog's suffering weigh any less in the moral balance of the cosmos?
>
> (Kohak, 1984, p. 92)

A commitment to animal alternatives is rightly considered a gauge of our scientific *and* moral advancement (Langley, 1989; Sharpe, 1988); social workers would be outraged, and rightly so, were it to be suggested that we might consider utilising the comatose, the senile and those with dementia, the irreversibly brain damaged, newborn but orphaned infants or anencephalic babies in the ways in which we currently utilise animals (who possess *at least* the capacities of the human beings mentioned – outrage is indeed selective) if by so doing it might be proved that it might benefit the health of the rest of *us*. The essence of inhumanity, Shaw (1934) reminds us, lies not in our hating our fellow creatures, but in our *indifference* to them.

A not insignificant factor in the reluctance to grant animals entrance into the circle of moral considerability is that it demands a revolution in thinking *and* practice in so many areas of our lives that we most likely take for granted – not least the claims that vegetarianism is morally obligatory (Preece, 2008; Walters and Portmess, 1999), as well as the health and environmental arguments supporting its adoption (Akers, 1989; Singer and Mason, 2006). It is an odd morality that accepts the subjectivity and consanguinity of our fellow creatures, and yet continues to treat them (or at least those who are not our *pets*) as objects fit for any purpose that we deem humanly advantageous.

If we return to a consideration of the five case studies in Chapter 1, it is as well to note that a significant factor in the reluctance of social workers to include non-human animals in their moral judgements (apart from the latter being seen as *exclusively* concerned with human welfare) rests upon their anxiety that to voice concern for the animals involved is to add further to the burdens under which these families and individuals are already labouring. But a social worker who overlooked the domestic violence directed against a woman, or the physical abuse or neglect visited upon her children by her husband and the children's father, on the grounds that his retrenchment and subsequent unemployment, and consequent depression, ought not to be compounded by adding to his woes, would be seen as acting improperly and indeed unethically. We would expect that the social worker working with this man would extend respect whilst also challenging those behaviours that violate the inherent value of the individuals who go to make up the family.

When social workers make the moral judgement (for that is what they invariably *do*) that they will *not* speak out about animal abuse or neglect, they routinely minimise this reality, and by inference deny that animals have any meaningful moral value. But as Moore (1992, p. 324) avers, '*All* beings are *ends*; *no* creatures are *means*. All beings have not equal rights, neither have all men; but *all have rights*.' Given that social work has a long and proud history of speaking out on behalf of, and of giving *moral priority*, to the weak and the vulnerable in human society, it is incumbent that social workers speak out and accord moral priority to animals, the most weak and vulnerable members of our communities. The social workers in each of the moral dilemmas posed can go a long way towards their resolution if they acknowledge that animals are our neighbours (and much more besides), and that morality is indubitably connected not only with attention to human individuals, but '*individual realities of other kinds ... towards the great surprising variety of the world*' (Murdoch, 1996, pp. 38, 66; emphasis mine).

All the animals in the five case studies are profoundly dependent upon their human companions, and the humans involved (especially the social workers) have a special responsibility for their welfare and well-being. And it is this dependency that demands human responsibility, imposing upon us acquired and positive duties, and upon which our claims to be moral agents either stand or fall. Social workers have a special responsibility to the weak and vulnerable of *all* species; once it is acknowledged that animals are part and parcel of the moral fabric, the sooner we shall come to see that we have duties and obligations that extend beyond the confines of our own species. Indeed social workers of a century hence may well come to view with incredulity the fact that their predecessors ever failed to extend respect to sentient creatures, or chose to remain morally indifferent to their plight. If we conceive reality as 'the whole of creation' (Wilkes, 1981, p. 106), we will extend respect to *all* individuals. It is not a being's intellectual capacity that entitles it to moral consideration, rather a creature's capacity for emotional fellowship (Midgley, 1985), whilst everyday, run-of-the-mill morality singles out the centrality of interests, not rationality (Sapontzis, 1987). If social workers so attend, they cannot go far wrong.

Social work is essentially concerned with, and engaged in, acquiring knowledge of the individual, knowledge understood as *loving union* (Midgley, 1995b), and Murdoch (1996, p. 30) reminds us that '*the* central concept of morality is "the individual" thought of as *knowable by love*' [emphasis mine]. The central concept of morality that ought to inform social work, and underpin its cardinal principle of *respect for individuals*, is *the individual, knowable by love, loving-kindness and loving union*.

As William James (Sapontzis, 1987, p. 110) entreats,

Take any demand, however slight, which any creature, however weak, may make. Ought it not, for its own sake, to be satisfied? If not, prove why not.

# Appendix – New Beginnings, Other Ends: An Inclusive Social Work Code of Ethics

> The emancipation of men from cruelty and injustice will bring with it in due course the emancipation of animals also. The two reforms are inseparably connected, and neither can be fully realized alone.
>
> *Henry Salt* (1921, p. 122).

The following Code is based upon the Australian Association of Social Workers Code of Ethics (1999b), but it articulates a substantive and inclusive revision.

## Purpose of Social Work

### Commitment and Aims

The discipline of social work is committed to the pursuit and maintenance of the well-being of the human and non-human animal, and the integrity of the natural world.

Social work aims to maximise the flourishing of human and non-human individuals, by way of attending to and respecting their needs and interests, and through an equal commitment to:

- working with and enabling human and non-human individuals to achieve the best possible levels of personal, social and species well-being; and
- working to achieve social and species justice through social development and social change, and attention to the moral value of individual subjects.

This involves:

- respecting and upholding the interests and rights of human and non-human individuals, and the natural world;
- working with individuals, groups and communities in the pursuit and achievement of equitable access to social, economic and political resources;
- providing assistance to enhance the well-being of the human and non-human animal, including entities such as individuals, families, groups, communities, organisations, societies and species, especially those who are neglected, vulnerable, disadvantaged or have exceptional needs;
- raising awareness of structural and species inequities;

166

- promoting policies and practices that achieve a fair and appropriate allocation of social resources respective to needs and interests of all species; and
- acting to bring about social and moral change to reduce social barriers, inequality, injustice and speciesism.

To accomplish its aims, the discipline of social work pursues:

- the development and application of knowledge, theory and skills regarding human and non-human animal behaviour, interests and needs, as well as social processes and social structures; and
- the development and redistribution of resources to meet the needs and maximise the flourishing of individuals, communities, species and the integrity of the natural world.

## Principles

In the determination and pursuit of its aims, social work is committed to the cardinal moral principle of *respect for individuals*, which entails:

- respect for the dignity and worth of human and non-human individuals;
- social justice and moral consideration for all human and non-human individuals;
- service to humanity, animality and the natural world;
- integrity in relation to moral practice towards human and non-human individuals; and
- competence in all actions towards human and non-human individuals.

In carrying out their disciplinary tasks and duties, social workers strive to act in ways that give equal priority to respect for the dignity and worth of all human and non-human individuals, and the pursuit of social and species justice. This commitment is demonstrated through service to human beings and non-human animals, integrity and competence, which characterise ethical social work practice.

Social work moral principles are derived from the principle of *respect for individuals*; together they underpin ethical social work practice.

## Value: Human and Non-Human Animal Inherent Value, Dignity and Worth

The discipline of social work holds that:

- every human and non-human individual has inherent value, dignity and worth, and is owed respect as a moral right; and
- each human and non-human individual has a right to well-being, self-fulfilment, self-determination and flourishing, consistent with the rights of others.

## Social workers' Principles

- respect the inherent value, dignity and worth of every human and non-human individual;
- respect the basic interests, needs and rights of human and non-human individuals, as well as the rights of families, groups, communities, societies and species, and the natural world;
- foster individual well-being, autonomy and personal/social responsibility, with due consideration for the rights of others; and
- recognise and respect group identity and interdependence and the collective needs of particular communities.

# Value: Social and Moral Justice

The discipline of social work holds that each society has a moral obligation to pursue social and moral justice, to provide maximum benefit for all its members, irrespective of species membership, and to afford them protection from harm.

The discipline understands social justice and moral consideration to encompass:

- the satisfaction of the basic needs and interests of human and non-human individuals;
- the equitable distribution of resources to meet these needs and interests and maximise flourishing;
- fair access to public services and benefits to achieve human and non-human potential and flourishing, and to respect the needs and interests of human and non-human individuals;
- recognition of individual, community, species and biosphere rights and duties;
- appropriate treatment and protection under the law, and equal moral consideration of interests; and
- social development in the interests of human and non-human animal welfare and well-being, and consistent with the integrity of the natural world.

## Principles

- promote distributive justice, social fairness and species justice, acting to reduce barriers and enhance the flourishing of all human and non-human individuals, with special regard for those who are disadvantaged, vulnerable, oppressed or have exceptional needs, irrespective of species membership;
- act to change social structures that preserve inequalities and injustice;
- meet their responsibilities to society and the natural world by engaging in action to: promote societal, species and biosphere well-being; advocate for equitable distribution of resources relative to needs and interests; and effect positive social and moral change in the interests of social and species justice, and the flourishing of human and non-human individuals;
- espouse the cause of human and non-human animal rights, affirming that civil and political rights must be accompanied by cultural, economic, moral, social and species rights;

- oppose and work to eliminate all violations of the rights, needs and interests of all human and non-human individuals;
- oppose prejudice and discrimination against any human or non-human individuals, and challenge views and actions that vilify, stereotype or render morally invisible human or non-human individuals;
- recognise and respect the racial, cultural and species diversity of society, taking into account the further diversity that exists among individuals, families, groups and communities within indigenous and other cultures, as well as other species;
- reject the abuse of power for exploitation or suppression; support policies and practices that aim to empower human beings, and respect the needs and interests of all human and non-human individuals;
- contribute disciplined knowledge and skill to aid individuals, groups, communities, societies and species in their development and in the resolution of conflicts and their consequences; and
- promote public participation in societal processes and decisions and in the development and implementation of social policies and services.

## Value: Service to Human and Non-Human Individuals

The social work discipline holds service in the interests of human and non-human well-being, and social and species justice, as primary objectives. The fundamental goals of social work service are to:

- attend to the interests and needs of human and non-human individuals; and
- enable human and non-human individuals to flourish.

### Principles

- in their practice, to place the objective of service before personal aims, views or advantage;
- to work with, on behalf of, or in the interests of all human and non-human individuals, to enable them to deal with personal and social difficulties and to obtain essential resources and services. This work may include, but is not limited to, interpersonal practice, groupwork, community work, social development, social action, policy development and research, and the development and nurturing of interdisciplinary linkages;
- in providing service, to apply their knowledge and skill in ways that maximise the benefit of their involvement;
- to recognise and respect individual and collective goals, responsibilities and differences;
- to be responsible for using their power and authority in ways that serve and respect human and non-human individuals, and the natural world; and
- to make morally and ethically accountable decisions based on their national and international codes of ethics, informed by the principle of *respect for individuals*.

# Bibliography

Ackroyd, Peter (1999) *Blake*. London: Vintage.

Adams, Carol (1990) *The Sexual Politics of Meat: A Feminist-Vegetarian Critical Theory*. Oxford: Polity Press.

—— (1994) *Neither Beast Nor Man: Feminism and the Defence of Animals*. New York: Continuum.

Adams, Carol and Donovan, Josephine (Eds) (1995) *Animals and Women: Feminist Theoretical Explorations*. Durham: Duke University Press.

Addams, Jane (1902) *Democracy and Social Ethics*. New York: Macmillan.

Akers, Keith (1989) *A Vegetarian Sourcebook: The Nutrition, Ecology, and Ethics of a Natural Foods Diet*. Arlington: Vegetarian Press.

Alexander, L. (1972) 'Social work's Freudian deluge: myth or reality?', *Social Services Review*, 46(4), pp. 517–38.

Almond, Brenda (2006) *The Fragmenting Family*. Oxford: Clarendon.

Altman, Nathaniel (1980) *Ahimsa: Dynamic Compassion*. Wheaton: Quest.

Anderson, Robert, Hart, Benjamin and Hart, Lynette (Eds) (1984) *The Pet Connection: Its Influence on Our Health and Quality of Life*. Minneapolis: Center to Study Human-Animal Relationships and Environments.

Annas, Julia (2003) *Plato*. Oxford: Oxford University Press.

Antonaccio, Maria (2000) *Picturing the Human: The Moral Thought of Iris Murdoch*. Oxford: Oxford University Press.

Appleyard, Brian (1998) *Brave New Worlds: Staying Human in the Genetic Future*. New York: Viking.

Aquinas, Thomas (1989) 'Differences between rational and other creatures', in Tom Regan and Peter Singer (Eds) *Animal Rights and Human Obligation* (Second Edition) Englewood Cliffs: Prentice-Hall, pp. 6–9.

Aristotle (1952) *Works of Aristotle: Volume 11*. Chicago: Encyclopaedia Britannica.

Armstrong, Edward (1973) *Saint Francis: Nature Mystic*. Berkeley: University of California Press.

Armstrong, John (2001) *The Intimate Philosophy of Art*. London: Penguin.

Ascione, Frank and Arkow, Phil (Eds) (1999) *Child Abuse, Domestic Violence, and Animal Abuse: Linking the Circles of Compassion for Prevention and Intervention*. West Lafayette: Purdue University Press.

Attfield, Robin (1983) 'Western traditions and environmental ethics', in Robert Elliott and Arran Gare (Eds) *Environmental Philosophy: A Collection of Readings*. St Lucia: University of Queensland Press, pp. 201–30.

Augustine, St. (1990) 'Rational domination', in Paul Clarke and Andrew Linzey (Eds) *Political Theory and Animal Rights*. London: Pluto Press, pp. 59–60.

Australian Association of Social Workers (1999a) *The Human-Animal Bond: Implications for Social Work*, Queensland Branch Newsletter, 4, December.

Australian Association of Social Workers (1999b) *Code of Ethics*. Kingston: AASW.

Baker, Steve (1993) *Picturing the Beast: Animals, Identity and Representation*. Manchester: Manchester University Press.

Banks, Sarah (1995) *Ethics and Values in Social Work*. London: Macmillan.

Barash, David (1980) *Sociobiology: The Whisperings Within*. London: Souvenir Press.

Barkow, Jerome, Cosmides, Lena and Tooby, John (Eds) (1992) *The Adapted Mind: Evolutionary Psychology and the Generation of Culture*. Oxford: Oxford University Press.

Barthes, Roland (1986) *The Rustle of Language*. New York: Hill & Wang.

Bateson, Gregory (1973) *Steps towards an Ecology of Mind*. London: Paladin.

Bateson, Patrick (1990) 'Choice, preference and selection', in Marc Bekoff and Dale Jamieson (Eds) *Interpretation and Explanation in the Study of Animal Behavior: Volume 1*. Boulder: Westview Press, pp. 149–56.

—— (2000) 'Taking the stink out of instinct', in Hilary and Steven Rose (Eds) *Alas, Poor Darwin: Arguments against Evolutionary Psychology*. London: Jonathan Cape, pp. 157–73.

Battersby, Christine (1980) 'Review of *Beast and Man*', *Philosophy*, 55, pp. 270–3.

Baum, Rainer (1988) 'Holocaust: moral indifference as *the* form of modern evil', in Alan Rosenberg and Norman Myers (Eds) *Echoes from the Holocaust: Philosophical Reflections on a Dark Time*. Philadelphia: Temple University Press, pp. 53–90.

Bauman, Clarence (1990) *The Sermon on the Mount: The Modern Quest for its Meaning*. Macon: Mercer University Press.

Bauman, Zygmunt (2000) *Modernity and the Holocaust*. Ithaca: Cornell University Press.

—— (2001) *The Individualized Society*. Cambridge: Polity.

Bavidge, Michael and Ground, Ian (1994) *Can We Understand Animal Minds?* London: Bristol Classical Press.

Beardsmore, Richard (1996) 'If a lion could talk', in Kjell Johannessen and T. Nordenstam (Eds) *Wittgenstein and the Philosophy of Culture*. Vienna: Lichler-Tempsky, pp. 41–59.

Beck, Alan and Katcher, Aaron (1996) *Between Pets and People: The Importance of Animal Companionship*. West Lafayette: Purdue University Press.

Bekoff, Marc (2002) *Minding Animals: Awareness, Emotions, and Heart*. New York: Oxford University Press.

Bekoff, Marc and Pierce, Jessica (2009) *Wild Justice: The Moral Lives of Animals*. Chicago: University of Chicago Press.

Benhabib, Seyla (1992) *Situating the Self: Gender, Community and Postmodernism in Contemporary Ethics*. New York: Routledge.

Benson, Thomas (1983) 'The clouded mirror: animal stereotypes and human cruelty', in Harlan Miller and William Williams (Eds) *Ethics and Animals*. Clifton: Humana Press, pp. 79–90.

Benton, Ted (1992) 'Animals and us: relations or ciphers?', *History of the Human Sciences*, 5(2), pp. 123–30.

—— (1993) *Natural Relations: Ecology, Animal Rights and Social Justice*. London: Verso.

Berlin, Isaiah (1986) *Four Essays on Liberty*. Oxford: Oxford University Press.

Bernstein, Saul (1975) 'Self-determination: king or citizen in the realm of values?', in F.E. McDermott (Ed.) *Self-Determination in Social Work*. London: Routledge & Kegan Paul, pp. 33–42.

Biddulph, Steve (2006) *Raising Babies: Should Under Threes Go to Nursery?* London: Harper Thorsons.

Biestek, Felix (1973) *The Casework Relationship.* London: Unwin University Books.

Birch, Charles (1995) *Feelings.* Sydney: University of New South Wales Press.

——— (1999) *Biology and the Riddle of Life.* Sydney: University of New South Wales Press.

Birke, Linda (1994) *Feminism, Animals and Science.* Buckingham: Open University Press.

Black, Edwin (2003) *War against the Weak: Eugenics and America's Campaign to Create a Master Race.* New York: Nation Books.

Blackmore, Susan (1999) *The Meme Machine.* Oxford: Oxford University Press.

Blake, William (1941) *Poetry and Prose of William Blake* (Edited by Geoffrey Keynes) London: Nonesuch Press.

Boakes, Robert (1984) *From Darwin to Behaviourism: Psychology and the Minds of Animals.* Cambridge: Cambridge University Press.

Boghossian, Paul (2006) *Fear of Knowledge: Against Relativism and Constructivism.* Oxford: Clarendon.

Bondurant, Joan (1965) *Conquest of Violence: The Gandhian Philosophy of Conflict.* Berkeley: University of California Press.

Booth, Charles (1889) *Life and Labour of the People of London.* London: Macmillan.

Bowlby, John (2005) *A Secure Base.* London: Routledge.

Bowpitt, Graham (1998) 'Evangelical Christianity, secular humanism, and the genesis of British social work', *British Journal of Social Work*, 28, pp. 675–93.

Braithwaite, Victoria (2010) *Do Fish Feel Pain?* Oxford: Oxford University Press.

British Union for the Abolition of Vivisection (2010), http://www.buav.org, accessed 25 June 2010.

Brophy, Brigid (1979) 'The Darwinist's dilemma', in David Paterson and Richard Ryder (Eds) *Animals' Rights: A Symposium.* Fontana: Centaur, pp. 63–72.

——— (1980) 'Animal happiness', *London Review of Books*, 2(11), pp. 4–6.

Browne, Annette (1995) 'The meaning of respect: a First Nations perspective', *The Canadian Journal of Nursing Research*, 27(4), pp. 95–110.

Buber, Martin (1970) *I and Thou.* Edinburgh: T. and T. Clark.

Budiansky, Stephen (1999) *If a Lion Could Talk: How Animals Think.* London: Phoenix.

Bustad, Leo (1980) *Animals, Ageing and the Aged.* Minneapolis: University of Minneapolis Press.

Butler, Joseph (1886) *The Analogy of Religion.* London: George Bell & Sons.

Butrym, Zofia (1982) *The Nature of Social Work.* London: Macmillan.

Callicott, J. Baird (1982) 'Traditional American Indian and Western European attitudes toward nature: an overview', *Environmental Ethics*, 4, pp. 293–318.

——— (1989) 'The metaphysical implications of ecology', in J. Baird Callicott (Ed.) *Nature in Asian Traditions of Thought: Essays in Environmental Philosophy.* New York: State University of New York Press, pp. 51–64.

——— (1992) 'Animal liberation and environmental ethics: back together again', in Eugene Hargrove (Ed.) *The Animal Rights/Environmental Debate: The Environmental Perspective.* Albany: State University of New York Press, pp. 249–59.

Canovan, Margaret (1977) *G.K. Chesterton: Radical Populist.* New York: Harcourt Brace Jovanovitch.

Carruthers, Peter (1992) *The Animals Issue: Moral Theory in Practice*. Cambridge: Cambridge University Press.

Cavalieri, Paola and Singer, Peter (Eds) (1993) *The Great Ape Project: Equality Beyond Humanity*. London: Fourth Estate.

Cave, George (1982) 'On the irreplaceability of animal life', *Ethics and Animals*, 3(4), pp. 106–16.

Chalmers, David (1995) 'The puzzle of consciousness', *Scientific American*, 273(6), pp. 80–6.

Chambon, Adrienne and Irving, Allan (Eds) (1994) *Essays in Postmodernism and Social Work*. Toronto: Canadian Scholars' Press.

Chapple, Christopher (1993) *Nonviolence to Animals, Earth and Self in Asian Traditions*. Albany: State University of New York Press.

—— (2002) (Ed.) *Jainism and Ecology: Nonviolence in the Web of Life*. Cambridge: Harvard University Press.

Chatterjee, Margaret (1985) *Gandhi's Religious Thought*. Basingstoke: Macmillan.

Chesterton, G.K. (1912) *What's Wrong with the World*. London: Cassell.

—— (1922) *Eugenics and Other Evils*. London: Cassell.

—— (1949) *Selected Essays of G.K. Chesterton*. London: Methuen.

Chomsky, Noam (1968) *Language and Mind*. New York: Harcourt, Brace & World.

Clark, Chris (2000) *Social Work Ethics: Politics, Principles and Practice*. Basingstoke: Palgrave Macmillan.

—— (2006) 'Moral character in social work', *British Journal of Social Work*, 36, pp. 75–89.

Clark, Chris and Asquith, Stewart (1985) *Social Work and Social Philosophy: A Guide for Practice*. London: Routledge & Kegan Paul.

Clark, Kenneth (1977) *Animals and Men*. New York: William Morrow.

Clark, Stephen R.L. (1975) *Aristotle's Man*. Oxford: Clarendon.

—— (1977) *The Moral Status of Animals*. Oxford: Clarendon.

—— (1978) 'Animal wrongs', *Analysis*, 38(3), pp. 147–9.

—— (1982) *The Nature of the Beast: Are Animals Moral?* Oxford: Oxford University Press.

—— (1983) 'Gaia and the forms of life', in Robert Elliot and Arran Gare (Eds) *Environmental Philosophy: A Collection of Readings*. St Lucia: University of Queensland Press, pp. 182–97.

—— (1984) *From Athens to Jerusalem: The Love of Wisdom and the Love of God*. Oxford: Clarendon.

—— (1985a) 'Good dogs and other animals', in Peter Singer (Ed.) *In Defence of Animals*. Oxford: Basil Blackwell, pp. 41–51.

—— (1985b) 'Animals in ethical tradition', in Norman Marsh and Susan Haywood (Ed.) *Animal Experimentation*. Nottingham: FRAME, pp. 1–6.

—— (1985c) 'The rights of the wild and the tame', *Chronicles of Culture*, 9(8), pp. 20–2.

—— (1986a) *The Mysteries of Religion*. Oxford: Basil Blackwell.

—— (1986b) 'With rationality and love', *Times Literary Supplement*, 26 September, pp. 1047–9.

—— (1989) 'How to reason about value judgments', in A. Phillips-Griffiths (Ed.) *Key Themes in Philosophy*. Cambridge: Cambridge University Press, pp. 173–90.

—— (1991) 'Eradicating the obvious', *Journal of Applied Philosophy*, 8(1), pp. 121–5.

—— (1993) *How to Think About the Earth: Philosophical and Theological Models for Ecology*. London: Mowbray.

—— (1995a) 'Status and contract societies: the nonhuman dimension', *National Geographical Journal of India*, 41, pp. 225–30.

—— (1995b) 'Objective values, final causes: Stoics, Epicureans, and Platonists', *Electronic Journal of Analytical Philosophy*, 3, Spring, pp. 1–7.

—— (1995c) 'Enlarging the community: companion animals', in Brenda Almond (Ed.) *Introducing Applied Ethics*. Oxford: Basil Blackwell, pp. 318–30.

—— (1996a) 'Commentary on Stephen Braude's "Multiple personality and moral responsibility" ', *Philosophy, Psychiatry and Psychology*, 3(1), March, pp. 55–8.

—— (1996b) 'Riots at Brightlingsea', *Journal of Applied Philosophy*, 13(1), pp. 109–12.

—— (1997) *Animals and Their Moral Standing*. London: Routledge.

—— (1998a) 'Objectivism and the alternatives', in Edgar Morscher and O. Neumaier and Peter Simons (Eds) *Applied Ethics in a Troubled World*. Dordrecht–Boston–London: Kluwer, pp. 285–94.

—— (1998b) 'Dangerous conservatives: a reply to Daniel Dombrowski', *Sophia*, 37(2), pp. 44–69.

—— (1998c) 'Understanding animals', in Michael Tobias & Kate Solisti-Mattelon (Eds) *Kinship with the Animals*. Hillsboro: Beyond Words, pp. 99–111.

—— (1998d) *God, Religion and Reality*. London: SPCK.

—— (1999) *The Political Animal: Biology, Ethics and Politics*. London: Routledge.

—— (2000a) *Biology and Christian Ethics*. Cambridge: Cambridge University Press.

—— (2000b) 'How to deal with animals', in Giles Legood (Ed.) *Veterinary Ethics*. London: Continuum, pp. 49–62.

—— (2001) 'The status of zygotes', *Philosopher's Magazine*, 13, pp. 42–3.

—— (2003) 'Non-personal minds', in Anthony O'Hear (Ed.) *Minds and Persons*. Cambridge: Cambridge University Press, pp. 185–209.

Clifton, Merritt (1991) 'Who helps the helpless child?', *Animals' Agenda*, December, pp. 42–5.

Coetzee, J.M. (1999) *The Lives of Animals*. Princeton: Princeton University Press.

—— (2000) *Disgrace*. London: Vintage.

Colloms, Brenda (1982) *Victorian Visionaries*. London: Constable.

Cordner, Christopher (2002) *Ethical Encounter: The Depth of Moral Meaning*. Basingstoke: Palgrave Macmillan.

Cottingham, John (1983) 'Neo-naturalism and its pitfalls', *Philosophy*, 58, pp. 455–70.

Coy, Jennie (1988) 'Animals' attitudes to people', in Tim Ingold (Ed.) *What is an Animal?* London: Unwin Hyman, pp. 77–83.

Crocker, D. (1984) 'Anthropomorphism: bad practice, honest prejudice', in Georgina Ferry (Ed.) *The Understanding of Animals*. Malden: Blackwell, pp. 304–13.

Daly, Martin & Wilson, Margo (1999) *The Truth About Cinderella: A Darwinian View of Parental Love*. New Haven: Yale University Press.

Danto, A.C. (1967) 'Persons', in Paul Edwards (Ed.) *The Encyclopedia of Philosophy: Volume 6*. New York: Macmillan, pp. 110–14.

Darwall, Stephen (1977/78) 'Two kinds of respect', *Ethics*, 88, pp. 36–49.

Darwin, Charles (1936) *The Origin of Species and the Descent of Man.* New York: Modern Library.
—— (1945) *On Humus and the Earthworm.* London: Faber & Faber.
—— (1965) *The Expressions of the Emotions in Man and Animals.* Chicago: University of Chicago Press.
Davidson, Arnold (1991) 'The horror of monsters', in James Sheehan and Morton Sosna (Eds) *The Boundaries of Humanity: Humans, Animals, Machines.* Berkeley: University of California Press, pp. 36–67.
Dawkins, Marian Stamp (1985) 'The scientific basis for suffering in animals', in Peter Singer (Ed.) *In Defence of Animals.* Oxford: Basil Blackwell, pp. 27–40.
—— (1998) *Through Our Eyes Only? The Search for Animal Consciousness.* Oxford: Oxford University Press.
Dawkins, Richard (1989) *The Selfish Gene.* Oxford: Oxford University Press.
—— (1992) 'Is God a computer virus?', *New Statesman*, 18 December, pp. 42–5.
DeGrazia, David (1996) *Taking Animals Seriously: Mental Life and Moral Status.* Cambridge: Cambridge University Press.
—— (2002) *Animal Rights: A Very Short Introduction.* Oxford: Oxford University Press.
Dennett, Daniel (1978) *Brainstorms: Philosophical Essays on Mind and Psychology.* Cambridge: MIT Press.
—— (1996) *Darwin's Dangerous Idea.* London: Penguin.
Dennis, Norman and Halsey, Albert (1988) *English Ethical Socialism: Thomas More to R.H. Tawney.* New York: Oxford University Press.
Descartes, Rene (1989) 'Animals are machines', in Tom Regan and Peter Singer (Eds) *Animal Rights and Human Obligations.* Englewood Cliffs: Prentice-Hall, pp. 13–19.
de Schweinitz, Karl (1943) *England's Road to Social Security.* Cranbury: A.S. Barnes.
Des Pres, Terrence (1976) *The Survivor: An Anatomy of Life in the Death Camps.* New York: Oxford University Press.
de Waal, Frans (1996a) *Peacemaking Among Primates.* Cambridge: Cambridge University Press.
—— (1996b) *Good Natured: The Origins of Right and Wrong in Humans and Other Animals.* Cambridge: Harvard University Press.
de Waal, Frans and Tyack, Peter (Eds) (2003) *Animal Social Complexity: Intelligence, Culture, and Individualized Societies.* Cambridge: Harvard University Press.
Dickens, Charles (1986) *The Complete Novels.* London: Octopus.
Dickin, Maria (1950) *The Cry of the Animal.* London: PDSA.
Dombrowski, Daniel (1988) *Hartshorne and the Metaphysics of Animal Rights.* Albany: State University of New York Press.
—— (1997) *Babies and Beasts: The Argument from Marginal Cases.* Urbana: University of Illinois Press.
—— (2000) *Not Even a Sparrow Falls: The Philosophy of Stephen R.L. Clark.* East Lansing: Michigan State University Press.
Donovan, Josephine (1996) 'Animal rights and feminist theory', in Josephine Donovan and Carol Adams (Eds) *Beyond Animal Rights: A Feminist Caring Ethic for the Treatment of Animals.* New York: Continuum, pp. 34–59.
Donovan, Josephine and Adams, Carol (Eds) (1996) *Beyond Animal Rights: A Feminist Caring Ethic for the Treatment of Animals.* New York: Continuum.

Dostoevsky, Fyodor (1952) *The Brothers Karamazov*. Chicago: Encyclopedia Britannica.

Downie, R.S and Telfer, Elizabeth (1969) *Respect for Persons*. London: George Allen & Unwin.

——— (1980) *Caring and Curing: A Philosophy of Medicine and Social Work*. London: Methuen.

Dunayer, Joan (2001) *Animal Equality: Language and Liberation*. Derwood: Ryce.

Dupre, John (1990) 'The mental life of nonhuman animals', in Marc Bekoff and Dale Jamieson (Eds) *Interpretation and Explanation in the Study of Animal Behavior: Volume 1*. Boulder: Westview Press, pp. 428–48.

Eagleton, Terry (1997) *The Illusions of Postmodernism*. Oxford: Blackwell.

——— (1998) *The Eagleton Reader*. (Edited by Stephen Regan) Oxford: Blackwell.

——— (2002) *Sweet Violence: The Idea of the Tragic*. Oxford: Blackwell.

——— (2009) *Reason, Faith, and Revolution: Reflections on the God Debate*. New Haven: Yale University Press.

Eibl-Eibesfeldt, Ireuaus (1971) *Love and Hate: On the Natural History of Basic Behaviour Patterns*. London: Methuen.

Eisnitz, Gail (1997) *Slaughterhouse*. Amherst: Prometheus.

Elliot, Robert (1984) 'Rawlsian justice and non-human animals', *Journal of Applied Philosophy*, 1(1), pp. 95–106.

——— (1992) 'Intrinsic value, environmental obligation and naturalness', *Monist*, 75, pp. 138–60.

Elliott, Lula Jean (1931) *Social Work Ethics*. New York: American Association of Social Workers.

Elsdon–Baker, Fern (2009) *The Selfish Genius: How Richard Dawkins Rewrote Darwin's Legacy*. London: Icon.

Elston, Mary Anne (1990) 'Women and anti-vivisection in Victorian England 1870–1900', in Nicolaas Rupke (Ed.) *Vivisection in Historical Perspective*. London: Routledge, pp. 259–94.

Emmet, Dorothy (1962) 'Ethics and the social worker', *British Journal of Psychiatric Social Work*, 6(4), pp. 165–72.

Engelhardt, H. Tristram (1986) *The Foundations of Bioethics*. Oxford: Oxford University Press.

English, Jane (1975) 'Abortion and the concept of a person', *Canadian Journal of Philosophy*, 5(2), pp. 233–43.

Evans, E.P. (1906) *The Criminal Prosecution and Capital Punishment of Animals*. New York: Dutton.

Feinberg, Joel (1986) 'Abortion', in Tom Regan (Ed.) *Matters of Life and Death: New Introductory Essays in Moral Philosophy*. New York: Random House, pp. 256–93.

Fiddes, Nick (1992) *Meat: A Natural Symbol*. London: Routledge.

Fisher, John (1990) 'The myth of anthropomorphism', in Marc Bekoff and Dale Jamieson (Eds) *Interpretation and Explanation in the Study of Animal Behavior: Volume 1*. Boulder: Westview Press, pp. 96–116.

Fodor, Jerry (1999) 'Not so clever Hans', *London Review of Books*, 21(3), pp. 12–13.

Foer, Jonathan (2010) *Eating Animals*. London: Penguin.

Fook, Jan (2002) *Social Work: Critical Theory and Practice*. London: Sage.

Fook, Jan and Pearce, Bob (Eds) (1999) *Transforming Social Work Practice: Postmodern Critical Perspectives*. London: Routledge.

Fossey, Dian (1983) *Gorillas in the Mist*. London: Hodder & Stoughton.

Foucault, Michel (1970) *The Order of Things: The Archaeology of the Human Services*. London: Tavistock.

—— (1983) *Beyond Structuralism and Hermeneutics*. Chicago: University of Chicago Press.

Fox, Warwick (1991) 'Self and world: a transpersonal, ecological approach', *Revision: Journal of Consciousness and Change*, 13(3), pp. 116–21.

—— (1995) *Towards a Transpersonal Ecology: Developing New Foundations for Environmentalism*. Dartington Totnes: Resurgence.

Francione, Gary (1995a) *Animals, Property and the Law*. Philadelphia: Temple University Press.

—— (1995b) 'Abortion and animal rights: are they comparable issues?', in Carol Adams and Josephine Donovan (Eds) *Animals and Women: Feminist Theoretical Explorations*. Durham: Duke University Press, pp. 149–59.

Frankel, Charles (1966) 'The moral framework of welfare', in John Morgan (Ed.) *Welfare and Wisdom*. Toronto: University of Toronto Press, pp. 147–64.

Frankl, Viktor (1965) *Man's Search for Meaning*. New York: Washington Square Press.

French, Richard (1975) *Antivivisection and Medical Science in Victorian Society*. Princeton: Princeton University Press.

Frey, R.G. (1980) *Interests and Rights: The Case against Animals*. Oxford: Clarendon.

—— (1987) 'Animal parts, human wholes: on the use of animals as a source of organs for human transplants', in James Humber and Robert Almeder (Eds) *Biomedical Ethics Reviews*. Clifton: Humana Press, pp. 89–107.

Friedlander, Albert (1994) *Riders towards the Dawn: From Holocaust to Hope*. New York: Continuum.

Friedmann, Erika and Thomas, Sue and Eddy, Timothy (2000) 'Companion animals and human health: physical and cardiovascular influences', in Anthony Podberscek, Elizabeth Paul and James Serpell (Eds) *Companion Animals and Us: Exploring the Relationships between People and Pets*. Cambridge: Cambridge University Press, pp. 125–42.

Fromm, Erich (1982) *Greatness and Limitations of Freud's Thought*. London: Abacus.

Fuller, B. (1949) 'The messes animals make in metaphysics', *Journal of Philosophy*, 26, pp. 829–38.

Gaita, Raimond (1991) *Good and Evil: An Absolute Conception*. Basingstoke: Macmillan.

—— (1999) *A Common Humanity: Thinking About Love and Truth and Justice*. Melbourne: Text.

—— (2002) *The Philosopher's Dog*. Melbourne: Text.

Gandhi, Mohandas (1984) *Sarvodaya: The Welfare of All*. Ahmedabad: Navajivan.

—— (1988) *The Moral Basis of Vegetarianism*. Ahmedabad: Navajivan.

Garner, Robert (2002) 'Political ideology and the legal status of animals', *Animal Law*, 8, pp. 77–91.

Garnett, A. Campbell (1969) 'Conscience and conscientiousness', in Joel Feinberg (Ed.) *Moral Concepts*. London: Oxford University Press, pp. 80–92.

Geldard, Richard (1995) *The Vision of Emerson*. Rockport: Element.

George, Don (Ed.) (2003) *The Kindness of Strangers*. Footscray: Lonely Planet.

George, Mother Hildegard (1999) 'The role of animals in the emotional and moral development of children', in Frank Ascione and Phil Arkow (Eds) *Child*

*Abuse, Domestic Violence, and Animal Abuse: Linking the Circles for Prevention and Intervention*. West Lafayette: Purdue University Press, pp. 380–92.

Geras, Norman (1998) *The Contract of Mutual Indifference: Political Philosophy after the Holocaust*. London: Verso.

Ghiselin, M.T. (1974) *The Economy of Nature and the Evolution of Sex*. Berkeley: University of California Press.

Godlovitch, Roslind (1972) 'Animals and morals', in Stanley Godlovitch, Roslind Godlovitch and John Harris (Eds) *Animals, Men and Morals: An Enquiry into the Maltreatment of Animals*. New York: Taplinger, pp. 156–72.

Goldstein, Howard (1973) *Social Work Practice: A Unitary Approach*. Columbia: University of South Carolina Press.

—— (1984) (Ed.) *Creative Change: A Cognitive Humanistic Approach to Social Work Practice*. New York: Tavistock.

Gompertz, Lewis (1992) *Moral Inquiries: On the Situation of Man and of Brutes*. Fontwell: Centaur.

Goodall, Jane (1990) *Through a Window: My Thirty Years with the Chimpanzees of Gombe*. Boston: Houghton Mifflin.

Goodpaster, Kenneth (1978) 'On being morally considerable', *Journal of Philosophy*, 75, pp. 308–25.

Goodwin, Brian (1988) 'Organisms and minds: the dialectics of the animal-human interface in biology', in Tim Ingold (Ed.) *What is an Animal?* London: Unwin Hyman, pp. 100–9.

—— (1995) *How the Leopard Changed its Spots: The Evolution of Complexity*. London: Phoenix.

Goodwin, Michele (2006) *Black Markets: The Supply and Demand of Body Parts*. Cambridge: Cambridge University Press.

Gopnik, Alison (2009) *The Philosophical Baby: What Children's Minds Tell Us about Truth, Love, and the Meaning of Life*. London: Bodley Head.

Gordon, Linda (1977) *Woman's Body, Woman's Right: Birth Control in America*. Harmondsworth: Penguin.

Gordon, William (1965) 'Knowledge and value: their distinction and relationship in clarifying social work practice', *Social Work*, July, pp. 32–9.

Gould, Peter (1988) *Early Green Politics: Back to Nature, Back to the Land, and Socialism in Britain 1880–1900*. Brighton: Harvester Press.

Grandin, Temple (2009) *Animals Make Us Human: Creating the Best Life for Animals*. Boston: Houghton Mifflin Harcourt.

Grant, Anne (1999) 'Resistance to the link at a domestic violence shelter', in Frank Ascione and Phil Arkow (Eds) *Child Abuse, Domestic Violence, and Animal Abuse: Linking the Circles of Compassion for Prevention and Intervention*. West Lafayette: Purdue University Press, pp. 159–67.

Gray, John (2002) *Straw Dogs: Thoughts on Humans and Other Animals*. London: Granta.

Gray, Mel and Stofberg, J. (2000) 'Respect for persons', *Australian Social Work*, 53, pp. 55–61.

Green, T.H. (1986) *Lectures on the Principles of Political Obligations and Other Writings*. (Edited by Paul Harris) Cambridge: Cambridge University Press.

Greene, Graham (1957) *The Quiet American*. London: Reprint Society.

——— (1982) *Monsignor Quixote*. London: Bodley Head.

Greer, Germaine (1971) *The Female Eunuch*. London: Paladin.

Griffin, Donald R. (1981) *The Question of Animal Awareness: Evolutionary Continuity of Mental Experience*. New York: Rockefeller University Press.

——— (1984) *Animal Thinking*. Cambridge: Harvard University Press.

——— (1992) *Animal Minds*. Chicago: University of Chicago Press.

Grimm, Harold (1970) *The Reformation Era: 1500–1650*. London: Macmillan.

Gutierrez, Gustavo (1988) *A Theology of Liberation: History, Politics and Salvation*. Maryknoll: Orbis.

Haille, Philip (1969) *The Paradox of Cruelty*. Middletown: Wesleyan University Press.

Halmos, Paul (1966) *The Faith of the Counsellors*. New York: Schocken.

Hardin, Garrett (1979) 'Lifeboat ethics: the case against helping the poor', in James Rachels (Ed.) *Moral Problems*. New York: Harper & Row, pp. 279–91.

Hardy, Thomas (1930) *The Later Years of Thomas Hardy: 1892–1928*. New York: Macmillan.

——— (1976) *The Complete Poems of Thomas Hardy*. (Edited by James Gibson) London: Macmillan.

Harmon, M. Judd (1964) *Political Thought: From Plato to the Present*. New York: McGraw Hill.

Harris, Errol (1968) 'Respect for persons', in Richard De George (Ed.) *Ethics and Society: Original Essays on Contemporary Moral Problems*. London: Macmillan, pp. 111–32.

Harris, John (1998) 'Four legs good, personhood better!', *Res Publica*, 4(1), pp. 51–8.

Harris, Paul (1991) 'The work of the imagination', in Andrew Whiten (Ed.) *Natural Theories of Mind*. Oxford: Basil Blackwell, pp. 283–304.

Harrison, Beverley (1983) *Our Right to Choose: Toward a New Ethic of Abortion*. Boston: Beacon Press.

Harrison, Brian (1967) 'Religion and recreation in nineteenth-century England', *Past and Present*, 38, pp. 98–125.

——— (1982) *Peaceable Kingdom*. Oxford: Oxford University Press.

Harrison, Ruth (1964) *Animal Machines*. London: Vincent Stuart.

Harwood, Dix (2002) *Love for Animals and How It Developed in Great Britain (1928)*. Lampeter: Edwin Mellen Press.

Haught, John (1990) 'Religious and cosmic homelessness: some environmental implications', in Charles Birch, William Eakin and Jay McDaniel (Eds) *Liberating Life: Contemporary Approaches to Ecological Theology*. Maryknoll: Orbis, pp. 159–81.

——— (2003) *Deeper than Darwin: The Prospect for Religion in the Age of Evolution*. Cambridge: Westview Press.

Hay, Peter (2002) *Main Currents in Western Environmental Thought*. Sydney: University of New South Wales Press.

Hearne, Vicki (1987) *Adam's Task: Calling Animals by Name*. New York: Knopf.

Himmelfarb, Gertrude (1991) *Poverty and Compassion: The Moral Imagination of the Late Victorians*. New York: Knopf.

——— (1994) *On Looking Into the Abyss: Untimely Thoughts on Culture and Society*. New York: Knopf.

—— (1995) *The De-Moralization of Society: From Victorian Virtues to Modern Values*. New York: Knopf.

Hoban, Russell (1977) *Turtle Diary*. London: Picador.

Hobbes, Thomas (1904) *Leviathan*. Cambridge: Cambridge University Press.

Hofstadter, Richard (1965) *Social Darwinism in American Thought*. New York: Braziller.

Hollis, Florence (1967) 'Principles and assumptions underlying casework practice', in Eileen Younghusband (Ed.) *Social Work and Social Values*. London: Allen & Unwin, pp. 22–38.

—— (1972) *Casework: A Psychosocial Therapy*. New York: Random House.

Hollis, Martin (1977) *Models of Man: Philosophical Thoughts on Social Action*. Cambridge: Cambridge University Press.

Hopkins, Gerard Manley (1967) *Hopkins: Selections*. (Edited by Graham Storey) London: Oxford University Press.

Horne, Michael (1987) *Values in Social Work*. Aldershot: Wildwood House.

Howe, David (1979) 'Agency function and social work principles', *British Journal of Social Work*, 9(1), pp. 29–47.

Howe, David (1987) *An Introduction to Social Work Theory: Making Sense in Practice*. Aldershot: Wildwood House.

Hughes, John (1984) 'Towards a moral philosophy for social work', *Social Thought*, 10, Spring, pp. 3–17.

Hull, David (1978) 'A matter of individuality', *Philosophy of Science*, 45, pp. 335–60.

Humane Research Australia (2010), http://www.humaneresearchaustralia.org.au/ statistics) – 'Statistics – animal use in research and teaching, Australia', accessed 25 June 2010.

Hume, David (1958) *A Treatise of Human Nature*. Oxford: Oxford University Press.

Humphrey, Nick (1976) 'The social function of intellect', in Patrick Bateson and Robert Hinde (Eds) *Growing Points in Ethology*. Cambridge: Cambridge University Press, pp. 303–17.

—— (1979) 'Nature's psychologists', in B. Josephson and V. Ramachandran (Eds) *Consciousness and the Physical World*. London: Pergamon Press, pp. 57–75.

Hunt, Leonard (1978) 'Social work and ideology', in Noel Timms and David Watson (Eds) *Philosophy in Social Work*. London: Routledge & Kegan Paul, pp. 7–25.

Hursthouse, Rosalind (2000) *Ethics, Humans and Animals*. London: Routledge.

Hutton, J.S. (1983) 'Animal abuse as a diagnostic approach in social work: a pilot study', in Aaron Katcher and Alan Beck (Eds) *New Perspectives on Our Lives with Companion Animals*. Philadelphia: University of Philadelphia Press, pp. 444–7.

Hyde, Walter (1915/1916) 'The prosecution and punishment of animals and life-less things in the Middle Ages and modern times', *University of Pennsylvania Law Review*, LXIV, pp. 696–730.

Ife, Jim (1999) 'Postmodernism, critical theory and social work', in Jan Fook and Bob Pearce (Eds) *Transforming Social Work Practice: Postmodern Critical Perspectives*. London: Routledge, pp. 211–23.

—— (2001) *Human Rights and Social Work: Towards Right-Based Practice*. Cambridge: Cambridge University Press.

Imre, Roberta Wells (1982) *Knowing and Caring: Philosophical Issues in Social Work*. Washington: University Press of America.

—— (1984) 'The nature of knowledge in social work', *Social Work*, 29(1), pp. 41–5.

Ingold, Tim (1988a) 'Introduction', in Tim Ingold (Ed.) *What is an Animal?* London: Unwin Hyman, pp. 1–16.

—— (1988b) 'The animal in the study of humanity', in Tim Ingold (Ed.) *What is an Animal?* London: Unwin Hyman, pp. 84–9.

Interlandi, Jeneen (2009) 'Not just urban legend', *Newsweek*, 19th January, http://www.newsweek.com/id/178873.

Iyer, Raghavan (1973) *The Moral and Political Thought of Mahatma Gandhi*. New York: Oxford University Press.

Jablonka, Eva and Lamb, Marion (2005) *Evolution in Four Dimensions: Genetic, Epigenetic, Behavioral, and Symbolic Variation in the History of Life*. Cambridge: MIT Press.

Jamieson, Dale (2002) *Morality's Progress: Essays on Humans, Other Animals, and the Rest of Nature*. Oxford: Clarendon.

Jeffrey, R.C. (1985) 'Animal interpretation', in Brian McLaughlin and Ernest Lepore (Eds) *Actions and Events: Perspectives on the Philosophy of Davidson*. Oxford: Basil Blackwell, pp. 481–7.

Jesudasan, Ignatius (1984) *A Gandhian Theology of Liberation*. Maryknoll: Orbis.

Johnson, Edward (1991) 'Carruthers on consciousness and moral status', *Between the Species*, 7(4), pp. 190–2.

Johnson, Lawrence (1991) *A Morally Deep World: An Essay on Moral Significance and Environmental Ethics*. Cambridge: Cambridge University Press.

Jones, David (2004) 'The art of philanthropy', *Eureka Street*, 14(7), pp. 38–9.

Jones, Gareth Stedman (1971) *Outcast London: A Study in the Relationship Between Classes in Victorian Society*. Oxford: Clarendon.

—— (2004) *An End to Poverty? A Historical Debate*. New York: Columbia University Press.

Kalof, Linda (2007) *Looking at Animals in Human History*. London: Reaktion.

Kant, Immanuel (1964) *The Groundwork of the Metaphysic of Morals*. London: Hutchinson University Library.

—— (1990) 'Duties to animals are indirect', Paul Clarke and Andrew Linzey (Eds) *Political Theory and Animal Rights*. London: Pluto Press, pp. 126–7.

Kean, Hilda (1998) *Animal Rights: Political and Social Change in Britain Since 1800*. London: Reaktion.

Keith-Lucas, Alan (1953) 'The political theory implicit in social casework theory', *American Political Science Review*, XLVII, pp. 1076–91.

—— (1992) 'A socially sanctioned profession?', in P. Nelson Reid and Philip Popple (Eds) *The Moral Purposes of Social Work: The Character and Intentions of a Profession*. Chicago: Nelson-Hall, pp. 51–70.

Kelch, Thomas (1998) 'Toward a non-property status for animals', *New York University Environmental Law Journal*, 6, pp. 531–85.

Keller, Evelyn Fox (1991) 'Language and ideology in evolutionary theory: reading cultural norms into natural law', in James Sheehan and Morton Sosna (Eds)

*The Boundaries of Humanity: Humans, Animals, Machines.* Berkeley: University of California Press, pp. 85–102.

Kellert, Stephen and Wilson, Edward (Eds) (1993) *The Biophilia Hypothesis.* Washington: Island Press.

Kendrick, Kevin (1992) 'Considerations of personhood in nursing research: an ethical perspective', Keith Soothill, Christine Henry and Kevin Kendrick (Eds) *Themes and Perspectives in Nursing.* London: Chapman & Hall, pp. 175–86.

Kennedy, John (1992) *The New Anthropomorphism.* Cambridge: Cambridge University Press.

Kohak, Erazim (1984) *The Embers and the Stars.* Chicago: University of Chicago Press.

Kropotkin, Peter (1990) 'Nature teaches mutual aid', in Paul Clarke and Andrew Linzey (Eds) *Political Theory and Animal Rights.* London: Pluto Press, pp. 88–90.

Lago, D., Connell, C. and Knight, B. (1985) 'The effects of animal companionship on older persons living at home', in Konrad Lorenz (Ed.) *The Human-Pet Relationship: Proceedings.* Vienna: IEMT, pp. 34–46.

Laing, Jacqueline (1997) 'Innocence and consequentialism: inconsistency, equivocation and contradiction in the philosophy of Peter Singer', in David Oderberg and Jacqueline Laing (Eds) *Human Lives: Critical Essays on Consequentialist Bioethics.* London: Macmillan, pp. 196–224.

Laing, R.D. (1967) *Politics of Experience and the Bird of Paradise.* London: Penguin.

Langley, Gill (Ed.) (1989) *Animal Experimentation: The Consensus Changes.* New York: Chapman & Hall.

Lansbury, Coral (1985) *The Old Brown Dog: Women, Workers, and Vivisection in Edwardian England.* Wisconsin: University of Wisconsin Press.

Lawson, Dominic (1995) 'A special kind of baby', *Sunday Telegraph*, 18th June.

Leahy, Michael (1994) *Against Liberation: Putting Animals in Perspective.* London: Routledge.

Leneman, Leah (1997) 'The awakened instinct: vegetarianism and the women's suffrage movement in Britain', *Women's History Review*, 6(2), pp. 271–87.

Levinas, Emmanuel (1990) *Difficult Freedom: Essays on Judaism.* Baltimore: John Hopkins University Press.

Levy, Charles (1973) 'The value base of social work', *Journal of Education for Social Work*, 9, pp. 34–42.

—— (1976) *Social Work Ethics.* New York: Human Science Press.

Levy, Neil (2002) *Moral Relativism: A Short Introduction.* Oxford: Oneworld.

Lewis, C.S. (1946) *The Abolition of Man.* London: Geoffrey Bles.

—— (1947) *Vivisection.* Boston: New England Anti-Vivisection Society.

Lewis, Thomas, Lannon, Richard and Amini, Fari (2001) *A General Theory of Love.* New York: Knopf.

Lewontin, Richard (1977) *Biology as a Social Weapon.* Minnesota: Science for the People Editorial Collective.

Lewontin, Richard, Rose, Steven and Kamin, Leon (1984) *Not in Our Genes: Biology, Ideology and Human Nature.* New York: Pantheon.

Li, Chien-hui (2000) 'A union of Christianity, humanity, and philanthropy: the Christian tradition and the prevention of cruelty to animals in nineteenth-century England', *Society and Animals*, 8(3), pp. 265–85.

Linden, Eugene (1976) *Apes, Men and Language.* New York: Penguin.

Linzey, Andrew (1987) *Christianity and the Rights of Animals.* London: SPCK.

—— (1994) *Animal Theology*. London: SCM.

—— (Ed.) (2009a) *The Link Between Animal Abuse and Human Violence*. Brighton: Sussex Academic Press.

—— (2009b) *Why Animal Suffering Matters*. Oxford: Oxford University Press.

Linzey, Andrew and Cohn-Sherbok, Dan (1997) *After Noah: Animals and the Liberation of Theology*. London: SPCK.

Linzey, Andrew and Yamamoto, Dorothy (Eds) (1998) *Animals on the Agenda: Questions About Animals for Theology and Ethics*. London: SCM Press.

Loar, Lynn (1999) ' "I'll only help you if you have two legs": or why human service professionals should pay attention to cases involving cruelty to animals', in Frank Ascione and Phil Arkow (Eds) *Child Abuse, Domestic Violence, and Animal Abuse: Linking the Circles of Compassion for Prevention and Intervention*. West Lafayette: Purdue University Press, pp. 120–36.

Loar, Lynn and White, Kenneth (1998) 'Connections drawn between child and animal victims of violence', in Randall Lockwood and Frank Ascione (Eds) *Cruelty to Animals and Interpersonal Violence: Readings in Research and Application*. West Lafayette: Purdue University Press, pp. 314–17.

Locke, John (1990) 'Cruelty is not natural', in Paul Clarke and Andrew Linzey (Eds) *Political Theory and Animal Rights*. London: Pluto Press, pp. 119–21.

Lockwood, Randall and Ascione, Frank (Eds) (1998) *Cruelty to Animals and Interpersonal Violence: Readings in Research and Application*. West Lafayette: Purdue University Press.

Lodrick, Deryck (1981) *Sacred Cows, Sacred Places: Origins and Survivals of Animal Homes in India*. Berkeley: University of California Press.

Lorenz, Konrad (1966) *On Aggression*. London: Methuen.

Lovelock, James (1979) *Gaia: A New Look at Life on Earth*. New York: Oxford University Press.

Lowenberg, Frank and Dolgoff, Ralph (1996) *Ethical Decisions for Social Work Practice*. Belmont: Wadsworth.

Lowry, Elizabeth (1999) 'Like a dog', *London Review of Books*, 21(20), pp. 1–12.

Lukes, Steven (1987) *Marxism and Morality*. Oxford: Oxford University Press.

Mabey, Richard (2005) *Nature Cure*. London: Chatto & Windus.

MacCunn, John (1911) *Ethics of Social Work*. London: Constable.

MacIntyre, Alasdair (1999) *Dependent Rational Animals: Why Human Beings Need the Virtues*. Chicago: Open Court.

MacIver, A. (1948) 'Ethics and the beetle', *Analysis*, 8, pp. 65–70.

Macklin, Ruth (1984) 'Personhood and the abortion debate', in Jay Garfield and Patricia Hennessey (Eds) *Abortion: Moral and Legal Perspectives*. Amherst: University of Massachusetts Press, pp. 81–102.

Maclagan, W.G. (1960a) 'Respect for a person as a moral principle – 1', *Philosophy*, 35(134), pp. 193–217.

—— (1960b) 'Respect for persons as a moral principle – 11', *Philosophy*, 35(135), pp. 289–305.

*Macquarie Dictionary* (1990) Sydney: Macquarie Library.

Malik, Kenan (2000) *Man, Beast and Zombie: What Science Can and Cannot Tell Us About Human Nature*. London: Weidenfeld & Nicolson.

Mandler, Peter (2004) 'Gold out of straw', *London Review of Books*, 26(4), p. 30.

Manning, Aubrey and Serpell, James (Eds) (1994) *Animals and Human Society: Changing Perspectives*. London: Routledge.

Marris, Peter (1974) *Loss and Change*. London: Routledge & Kegan Paul.

Marston, Greg and Watts, Rob (2004) 'The problem with neo-conservative social policy: rethinking the ethics of citizenship and the welfare state', *Just Policy*, 33, pp. 34–45.

Marx, Karl (1990) 'An animal is not a species being', in Paul Clarke and Andrew Linzey (Eds) *Political Theory and Animal Rights*. London: Pluto Press, pp. 42–4.

Mascaro, Juan (1970) *The Bhagavad Gita*. London: Rider.

Maslow, Abraham (1968) *Towards a Psychology of Being*. Princeton: D. Van Nostrand.

—— (1993) *The Farther Reaches of Human Nature*. New York: Penguin/Arkana.

Mason, Jim (1993) *An Unnatural Order: Uncovering the Roots of Our Domination of Nature and Each Other*. New York: Simon & Schuster.

Mason, Jim and Singer, Peter (1980) *Animal Factories*. New York: Crown.

Masson, Jeffrey (2003) *The Pig Who Sang to the Moon: The Emotional World of Farm Animals*. New York: Ballantine.

Masson, Jeffrey and McCarthy, Susan (1996) *When Elephants Weep: The Emotional Lives of Animals*. London: Vintage.

McCormack, John (2001) 'How many social workers now? A review of census and other data', *Australian Social Work*, 54(3), pp. 63–72.

McDaniel, Jay (1989) *Of God and Pelicans: A Theology of Reverence for Life*. Louisville: Westminster/John Knox Press.

McDermott, F.E. (1975) 'Introduction', in F.E. McDermott (Ed.) *Self-Determination in Social Work*. London: Routledge & Kegan Paul, pp. 1–14.

McKinnon, Susan (2006) *Neo-Liberal Genetics: The Myths and Moral Tales of Evolutionary Psychology*. Chicago: Prickly Paradigm Press.

Medawar, Peter (1957) *The Uniqueness of the Individual*. London: Methuen.

Meemeduma, Pauline and Atkinson, Rachel (1996) 'The way forward – culturally relevant practice responses to child abuse in Aboriginal communities in North Queensland', in *Proceedings from Australasian Child Abuse and Neglect Conference*, Melbourne.

Melden, A.I. (1988) *Rights in Moral Lives: A Historical-Philosophical Essay*. Berkeley: University of California Press.

Merchant, Carolyn (1990) *The Death of Nature: Women, Ecology and the Scientific Revolution*. San Francisco: Harper Collins.

Midgley, Mary (1973) 'The concept of beastliness: philosophy, ethics and animal behaviour', *Philosophy*, 48, pp. 111–35.

—— (1980) 'The lack of gap between fact and value', *Proceedings of the Aristotelian Society*, Supplement 54, pp. 207–23.

—— (1983a) *Animals and Why They Matter*. Athens, Georgia: University of Georgia Press.

—— (1983b) *Heart and Mind: The Varieties of Moral Experience*. London: Methuen.

—— (1983c) 'Duties concerning islands', in Robert Elliot and Arran Gare (Eds) *Environmental Philosophy: A Collection of Readings*. St Lucia: University of Queensland Press, pp. 166–81.

—— (1983d) 'Human ideals and human needs', *Philosophy*, 58, pp. 89–94.

—— (1984a) 'De-dramatizing Darwin', *Monist*, 67, pp. 200–15.

—— (1984b) 'On being terrestrial', *Philosophy*, Supplement, pp. 79–91.

—— (1985) 'Persons and non-persons', in Peter Singer (Ed.) *In Defence of Animals*. Oxford: Basil Blackwell, pp. 52–62.

—— (1986) *Evolution as a Religion: Strange Hopes and Stranger Fears*. London: Methuen.

—— (1989) 'Practical solutions', *Hastings Center Report*, 19, pp. 44–5.

—— (1990) 'The use and uselessness of learning', *The European Journal of Education*, 25(3), pp. 283–94.

—— (1991) 'The origin of ethics', in Peter Singer (Ed.) *A Companion to Ethics*. Oxford: Basil Blackwell, pp. 3–13.

—— (1992) 'Is the biosphere a luxury?', *Hastings Center Report*, 22(3), pp. 7–12.

—— (1993) *Can't We Make Moral Judgements?* New York: St Martin's Press.

—— (1994a) 'Darwinism and ethics', in K. Fulford, Grant Gillett and Janet Soskice (Eds) *Medicine and Moral Reasoning*. Cambridge: Cambridge University Press, pp. 6–18.

—— (1994b) 'Bridge-building at last', in Aubrey Manning and James Serpell (Eds) *Animals and Human Society: Changing Perspectives*. London: Routledge, pp. 188–94.

—— (1995a) *The Ethical Primate: Humans, Freedom, Morality*. London: Routledge.

—— (1995b) *Wisdom, Information and Wonder: What is Knowledge For?* London: Routledge.

—— (1995c) 'The ethical primate', *Journal of Consciousness Studies*, 2(1), pp. 67–75.

—— (1996a) *Beast and Man: The Roots of Human Nature*. London: Routledge.

—— (1996b) *Wickedness: A Philosophical Essay*. London: Routledge.

—— (1996c) *Utopias, Dolphins and Computers: Problems of Philosophical Plumbing*. London: Routledge.

—— (1999) 'Being scientific about ourselves', *Journal of Consciousness Studies*, 6, April, pp. 85–98.

—— (2001) *Science and Poetry*. London: Routledge.

—— (2002) 'Pluralism: the many maps model', *Philosophy Now*, 35 (March/April), pp. 10–11.

—— (2003) *The Myths We Live By*. London: Routledge.

—— (2010) *The Solitary Self: Darwin and the Selfish Gene*. Durham: Acumen.

Midgley, Mary and Hughes, Judith (1983) *Women's Choices: Philosophical Problems Facing Feminism*. London: Weidenfeld & Nicolson.

Miles, Arthur (1954) *American Social Work Theory: A Critique and a Proposal*. New York: Harper.

Mill, John Stuart (1901) *Utilitarianism*. London: Longmans, Green & Co.

Millard, David (1977) 'Literature and the therapeutic imagination', *British Journal of Social Work*, 7(2), pp. 173–84.

Miller, Henry (1968) 'Value dilemmas in social casework', *Social Work*, 13(1), pp. 27–33.

Moffett, Jonathan (1968) *Concepts of Casework Treatment*. London: Routledge & Kegan Paul.

Moore, George E. (1903) *Principia Ethica*. Cambridge: Cambridge University Press.

Moore, J. Howard (1992) *The Universal Kinship*. Fontwell: Centaur.

Morley, Louise and Ife, Jim (2002) 'Social work and a love of humanity', *Australian Social Work*, 55(1), pp. 69–77.

Morris, Helbert (1968) 'Persons and punishment', *Monist*, 52, pp. 475–501.

Moss, Arthur (1961) *Valiant Crusade: The History of the RSPCA*. London: Cassell.

Murdoch, Iris (1977) *The Fire and the Sun: Why Plato Banished the Artists*. Oxford: Oxford University Press.

——— (1987) *Acastos: Two Platonic Dialogues*. New York: Viking.

——— (1988) *The Philosopher's Pupil*. London: Penguin.

——— (1993) *Metaphysics as a Guide to Morals*. London: Penguin.

——— (1996) *The Sovereignty of Good*. London: Routledge.

——— (1997) *Existentialists and Mystics: Writings in Philosophy and Literature*. (Edited by Peter Conradi) London: Chatto & Windus.

Naess, Arne (1997) 'Heidegger, postmodern theory and deep ecology', *Trumpeter: Journal of Ecosophy*, 14, pp. 181–3.

Nagel, Thomas (1970) *The Possibility of Altruism*. Princeton: Princeton University Press.

——— (1974) 'What is it like to be a bat?', *Philosophical Review*, 83, pp. 435–50.

Nash, Roderick (1990) *The Rights of Nature: A History of Environmental Values*. Leichhardt: Primavera Press/Wilderness Society.

Nelkin, Dorothy (1999) 'Behavioral genetics and dismantling the welfare state', in Ronald Carson and Mark Rothstein (Eds) *Behavioral Genetics*. Baltimore: John Hopkins University Press, pp. 156–71.

——— (2000) 'Less selfish than sacred? Genes and the religious impulse in evolutionary psychology', in Hilary Rose and Steven Rose (Eds) *Alas Poor Darwin: Arguments Against Evolutionary Psychology*. London: Jonathan Cape, pp. 14–27.

Nelson, James (1988) 'Animals, handicapped children and the tragedy of marginal cases', *Journal of Medical Ethics*, 14, pp. 191–3.

Niebuhr, Reinhold (1932) *The Contribution of Religion to Social Work*. New York: Columbia University Press.

Noddings, Nel (1984) *A Feminist Approach to Ethics and Moral Education*. Berkeley: University of California Press.

Nozick, Robert (1995) *Anarchy, State and Utopia*. Oxford: Basil Blackwell.

Nussbaum, Martha (1992) 'Human functioning and social justice: in defence of Aristotelian essentialism', *Political Theory*, 20(2), pp. 202–46.

——— (2006) *Frontiers of Justice: Disability, Nationality, Species Membership*. Cambridge: Belknap Press.

O'Connor, Brendon (2001) 'The intellectual origins of welfare dependency', *Australian Journal of Social Issues*, 36(3), pp. 221–36.

O'Hear, Anthony (1997) *Beyond Evolution: Human Nature and the Limits of Evolutionary Explanations*. Oxford: Clarendon.

O'Neill, Onora (2002) *A Question of Trust*. Cambridge: Cambridge University Press.

Orwell, George (1984) *The Collected Essays, Journalism and Letters of George Orwell: Volume 111*. (Edited by Sonia Orwell and Ian Angus) Harmondsworth: Penguin/Martin Secker & Warburg.

——— (1993) *Animal Farm*. London: Compact Books.

Page, George (2000) *The Singing Gorilla: Understanding Animal Intelligence*. London: Headline.

Parfit, Derek (1976) 'Rights, interests, and possible people', in Samuel Gorovitz (Ed.) *Moral Problems in Medicine*. New York: Prentice-Hall, pp. 369–75.

Parton, Nigel and Marshall, Wendy (1998) 'Postmodernism and discourse approaches to social work', in Robert Adams, Lena Dominelli and Malcolm Payne (Eds) *Social Work: Themes, Issues and Critical Debates*. Basingstoke: Macmillan, pp. 240–9.

Parton, Nigel and O'Byrne, Patrick (2000) *Constructive Social Work: Towards a New Practice*. Basingstoke: Palgrave Macmillan.

Passmore, John (1974) *Man's Responsibility for Nature: Ecological Problems and Western Traditions*. London: Duckworth.

Patterson, Charles (2002) *Eternal Treblinka: Our Treatment of Animals and the Holocaust*. New York: Lantern.

Paul, Elizabeth (2000) 'Love of pets and love of people', in Anthony Podberscek, Elizabeth Paul and James Serpell (Eds) *Companion Animals and Us: Exploring the Relationship Between People and Pets*. Cambridge: Cambridge University Press, pp. 168–86.

Paul, Elizabeth and Serpell, James (1993) 'Childhood pet keeping and humane attitudes in young adulthood', *Animal Welfare*, 2, pp. 321–37.

Payne, Malcolm (1997) *Modern Social Work Theory*. Basingstoke: Macmillan.

Pearson, Geoffrey (1975) *The Deviant Imagination: Psychiatry, Social Work and Social Change*. London: Macmillan.

Peel, Mark (2003) *The Lowest Rung: Voices of Australian Poverty*. Cambridge: Cambridge University Press.

Peile, Colin (1993) 'Determinism versus creativity: which way for social work?', *Social Work*, 38(2), pp. 127–34.

PDSA (2010) http://www.pdsa.org.uk/mariadickin.html, accessed 25 June 2010.

Perlman, Helen Harris (1979) *Relationship: The Heart of Helping People*. Chicago: University of Chicago Press.

Pernick, Martin (1985) *A Calculus of Suffering: Pain, Professionalism and Anaesthesia in Nineteenth Century America*. New York: Columbia University Press.

Petchesky, Rosalind (1984) *Abortion and Women's Choice: The State, Sexuality, and Reproductive Freedom*. New York: Longman.

Peterson, Dale and Goodall, Jane (1993) *Visions of Caliban: On Chimpanzees and People*.

Phillips, Adam and Taylor, Barbara (2009) *On Kindness*. London: Penguin.

Philp, Mark (1979) 'Notes on the form of knowledge in social work', *Sociological Review*, 27, pp. 83–111.

Phipps, William (2002) *Darwin's Religious Odyssey*. Harrisburg: Trinity Press International.

Pinches, Charles and McDaniel, Jay B. (Eds) (1993) *Good News for Animals? Christian Approaches to Animal Well-Being*. Maryknoll: Orbis.

Pinker, Robert (1971) *Social Theory and Social Policy*. London: Heinemann.

Plant, Raymond (1970) *Social and Moral Theory in Casework*. London: Routledge & Kegan Paul.

Plant, Raymond, Lesser, Harry and Taylor-Gooby, Peter (1980) *Political Philosophy and Social Welfare: Essays on the Normative Basis of Welfare Provision*. London: Routledge & Kegan Paul.

Pluhar, Evelyn (1987) 'The personhood view and the argument from marginal cases', *Philosophica*, 39(1), pp. 23–8.

——— (1995) *Beyond Prejudice: The Moral Significance of Human and Nonhuman Animals*. Durham: Duke University Press.

Plumwood, Val (1992) 'Sealskin', *Meanjin*, 1, pp. 45–57.

——— (1997) '*Babe*: the tale of the speaking meat', *Animal Issues: Philosophical and Ethical Issues Related to Human/Animal Interactions*, 1(1), pp. 21–36.

Podberscek, Anthony, Paul, Elizabeth and Serpell, James (Eds) (2000) *Companion Animals and Us: Exploring the Relationships Between People and Pets*. Cambridge: Cambridge University Press.

Preece, Gordon (2002) 'The unthinkable and unlivable singer', in Gordon Preece (Ed.) *Rethinking Peter Singer: A Christian Critique*. Downers Grove: InterVarsity Press, pp. 23–67.

Preece, Rod (2005) *Brute Souls, Happy Beasts and Evolution: The Historical Status of Animals*. Vancouver: University of British Columbia Press.

——— (2008) *Sins of the Flesh: A History of Ethical Vegetarian Thought*. Vancouver: University of British Columbia Press.

Primatt, Humphry (1992) *The Duty of Mercy: And the Sin of Cruelty to Brute Animals*. Fontwell: Centaur.

Pumphrey, Muriel (1961) 'Transmitting values and ethics through social work practice', *Social Work*, 6, pp. 68–75.

Quackenbush, Jamie (1981) 'Social work in a veterinary hospital: a response to owner grief reactions', *Archives of the Foundation of Thanatology*, 9, p. 56.

Quinton, Anthony (1973) *The Nature of Things*. London: Routledge & Kegan Paul.

Rachels, James (1978) 'What people deserve', in John Arthur and William Shaw (Eds) *Justice and Economic Distribution*. Englewood Cliffs: Prentice-Hall, pp. 150–63.

——— (1995) *The Elements of Moral Philosophy*. New York: McGraw-Hill.

——— (1999) *Created from Animals: The Moral Implications of Darwinism*. Oxford: Oxford University Press.

Radner, Daisie and Radner, Michael (1996) *Animal Consciousness*. Amherst: Prometheus.

Ragg, Nicholas (1977) *People Not Cases: A Philosophical Approach to Social Work*. London: Routledge & Kegan Paul.

——— (1980) 'Respect for persons and social work: social work as "doing philosophy"', in Noel Timms (Ed.) *Social Welfare: Why and How?* London: Routledge & Kegan Paul, pp. 211–32.

Rawls, John (1999) *A Theory of Justice*. Oxford: Oxford University Press.

Reamer, Frederic (1995) *Social Work Values and Ethics*. New York: Columbia University Press.

Regan, Tom (1982) *All That Dwell Therein: Essays on Animal Rights and Environmental Ethics*. Berkeley: University of California Press.

——— (1983) *The Case for Animal Rights*. Berkeley: University of California Press.

——— (1989) 'Ill-gotten gains', in Gill Langley (Ed.) *Animal Experimentation: The Consensus Changes*. New York: Chapman & Hall, pp. 19–41.

——— (1991) *The Thee Generation*. Philadelphia: Temple University Press.

——— (2004) *Empty Cages: Facing the Challenge of Animal Rights*. Lanham: Rowman & Littlefield.

Reid, P. Nelson and Popple, Philip (Eds) (1992) *The Moral Purposes of Social Work: The Characters and Intentions of a Profession*. Chicago: Nelson-Hall.

Reist, Melinda Tankard (2000) *Giving Sorrow Words: Women's Stories of Grief after Abortion.* Potts Point: Duffy & Snellgrove.

———— (2006) *Defiant Birth: Women Who Resist Medical Eugenics.* North Melbourne: Spinifex Press.

Richmond, Mary (1922) *What is Social Casework?* New York: Russell Sage.

Ridley, Matt (1996) *The Origins of Virtue.* London: Viking.

Rips, Lance and Conrad, Frederick (1989) 'Folk psychology of mental activities', *Psychological Review,* 96(2), pp. 187–207.

Risley-Curtiss, Christina (2009) 'The role of animals in public child welfare work', in Andrew Linzey (Ed.) *The Link Between Animal Abuse and Human Violence.* Brighton: Sussex Academic Press, pp. 126–41.

Robinson, Virginia (1930) *A Changing Psychology in Social Casework.* Chapel Hill: University of North Carolina Press.

Rodd, Rosemary (1990) *Biology, Ethics and Animals.* Oxford: Clarendon.

———— (1996) 'Evolutionary ethics and the status of non-human animals', *Journal of Applied Philosophy,* 13(1), pp. 63–72.

Rogers, Carl (1951) *Client-Centered Therapy: Its Current Practice, Implications and Theory.* London: Constable.

Rogers, Lesley (1997) *Minds of Their Own: Thinking and Awareness in Animals.* St Leonards: Allen & Unwin.

Rollin, Bernard E. (1990) *The Unheeded Cry: Animal Consciousness, Animal Pain and Science.* Oxford: Oxford University Press.

———— (1995) *The Frankenstein Syndrome: Ethical and Social Issues in the Genetic Engineering of Animals.* Cambridge: Cambridge University Press.

Rose, Hilary (2000) 'Colonising the social sciences', in Hilary Rose and Steven Rose (Eds) *Alas, Poor Darwin: Arguments Against Evolutionary Psychology.* London: Jonathan Cape, pp. 106–28.

Rose, Hilary and Rose, Steven (1982) 'Moving Right out of welfare – and the way back', *Critical Social Policy,* 2, pp. 7–18.

Rose, Hilary and Rose, Steven (Eds) (2000a) *Alas, Poor Darwin: Arguments Against Evolutionary Psychology.* London: Jonathan Cape.

Rose, Hilary and Rose, Steven (2000b) 'Introduction', in Hilary and Steven Rose (Eds) *Alas, Poor Darwin: Arguments Against Evolutionary Psychology.* London: Jonathan Cape, pp. 1–13.

Rose, Steven (1997) *Lifelines: Biology, Freedom, Determinism.* London: Allen Lane/Penguin.

———— (1998) 'The genetics of blame', *New Internationalist,* April, pp. 20–1.

Roseberry, Kelly and Rovin, Laurie (1999) 'Animal-assisted therapy for sexually abused adolescent females', in Frank Ascione and Phil Arkow (Eds) *Child Abuse, Domestic Violence, and Animal Abuse: Linking the Circles of Compassion for Prevention and Intervention.* West Lafayette: Purdue University Press, pp. 433–42.

Rosenfield, Leonora Cohen (1941) *From Beast-Machine to Man-Machine.* New York: Oxford University Press.

Rousseau, Jean Jacques (1950) *The Social Contract and Discourses.* New York: Dutton.

Rowe, William (1996) 'Client-centered theory: a person-centered approach', in Francis J. Turner (Ed.) *Social Work Treatment: Interlocking Theoretical Approaches.* New York: Free Press, pp. 69–93.

Rowlands, Mark (1998) *Animal Rights: A Philosophical Defence*. Basingstoke: Macmillan.
—— (2002) *Animals Like Us*. London: Verso.
—— (2008) *The Philosopher and the Wolf: Lessons from the Wild on Love, Death and Happiness*. London: Granta.
RSPCA (2008–09) 'RSPCA Australia National Statistics 2008–2009', http://www.Rspca.org.au/assets/files/Resources/RSPCA, Annual Stats 2008–2009.pdf.
Rudman, Stanley (1997) *Concepts of Person and Christian Ethics*. Cambridge: Cambridge University Press.
Ruesch, Hans (1983) *Slaughter of the Innocent*. Zurich: Civis.
Ruse, Michael and Wilson, E.O. (1986) 'Moral philosophy as applied science', *Philosophy*, 61, pp. 173–92.
Ruskin, John (1900) *Unto This Last*. London: George Allen.
—— (1995) *Selected Writings*. London: J.M. Dent.
Ryan, Alan (1974) 'The nature of human nature in Hobbes and Rousseau', in Jonathan Benthall (Ed.) *The Limits of Human Nature*. New York: Dutton, pp. 3–19.
Ryan, Thomas David Anthony (1993a) *The Widening Circle: Should Social Work Concern Itself With Nonhuman Animal Rights?* (Unpublished Social Work Honours Thesis).
—— (1993b) 'Social work and nonhuman animal rights', *Northern Radius*, November, pp. 24–5.
Ryder, Richard D. (1983) *Victims of Science: The Use of Animals in Research*. London: National Anti-Vivisection Society.
—— (1989) *Animal Revolution: Changing Attitudes towards Speciesism*. Oxford: Basil Blackwell.
Salisbury, Joyce (1994) *The Beast Within: Animals in the Middle Ages*. New York: Routledge.
Salt, Henry S. (1900) 'The rights of animals', *International Journal of Ethics*, 10, pp. 206–22.
—— (1921) *Seventy Years Among Savages*. London: George Allen & Unwin.
—— (1935) *The Creed of Kinship*. London: Constable.
—— (1980) *Animals' Rights: Considered in Relation to Social Progress*. Clarks Summit: Society for Animal Rights.
Sapontzis, Steve (1987) *Morals, Reason and Animals*. Philadelphia: Temple University Press.
Sartre, Jean-Paul (1957) *Being and Nothingness: An Essay on Phenomenonological Ontology*. London: Methuen.
—— (1958) *Existentialism and Humanism*. London: Eyre Methuen.
Saunders, Peter (2004) *Australia's Welfare Habit and How to Kick It*. Melbourne: Duffy & Snellgrove.
Scarlett, Brain (1997) 'The moral uniqueness of the human animal', in David Oderberg and Jacqueline Laing (Eds) *Human Lives: Critical Essays on Consequentialist Bioethics*. London: Macmillan, pp. 77–95.
Schleifer, Harriet (1985) 'Images of death and life: food animal production and the vegetarian option', in Peter Singer (Ed.) *In Defence of Animals*. Oxford: Basil Blackwell, pp. 63–73.
Schwartz, Wynn (1982) 'The problem of other possible persons: dolphins, primates, and aliens', in Keith Davis and Thomas Mitchell (Eds) *Advances in Descriptive Psychology: Volume 2*. Greenwich: JAI Press, pp. 31–55.

Schweitzer, Albert (1955) *Civilization and Ethics*. London: Adam & Charles Black.

Scott, Dorothy and Swain, Shurlee (2002) *Confronting Cruelty: Historical Perspectives on Child Protection in Australia*. Melbourne: Melbourne University Press.

Searle, John (1983) *Intentionality*. Cambridge: Cambridge University Press.

―――― (1992) *The Rediscovery of Mind*. Cambridge: MIT Press.

Seidler, Michael (1977) 'Hume and the animals', *Southern Journal of Philosophy*, 15, pp. 361–72.

Senchuk, Dennis (1991) *Against Instinct*. Philadelphia: Temple University Press.

Sennett, Richard (1999) *The Corrosion of Character*. New York: W.W. Norton.

Serpell, James (1986) *In the Company of Animals: A Study of Human-Animal Relationships*. Oxford: Basil Blackwell.

Sessions, George (1995) 'Postmodernism and environmental justice', *Trumpeter: Journal of Ecosophy*, 12, pp. 150–4.

Shardlow, Steven (Ed.) (1989) *The Values of Change in Social Work*. London: Routledge.

Sharpe, Lynne (2005) *Creatures Like Us? A Relational Approach to the Moral Status of Animals*. Exeter: Imprint Academic.

Sharpe, Robert (1988) *The Cruel Deception: The Use of Animals in Medical Research*. Wellingborough: Thorsons.

Shaw, George Bernard (1928) *The Doctor's Dilemma*. London: Constable.

―――― (1934) 'The devil's disciple', in *The Complete Plays*. London: Oldhams Press, pp. 218–50.

―――― (1949) *Shaw on Vivisection*. London: George Allen & Unwin.

Shaw, John (1974) *The Self in Social Work*. London: Routledge & Kegan Paul.

Sheehan, James (1991) 'Introduction', in James Sheehan and Morton Sosna (Eds) *The Boundaries of Humanity: Humans, Animals, Machines*. Berkeley: University of California Press, pp. 27–35.

Sheldrake, Rupert (1991) *The Rebirth of Nature: The Greening of Science and God*. London: Rider.

―――― (2000) *Dogs That Know When Their Owners Are Coming Home and Other Unexplained Powers of Animals*. London: Arrow.

Simpkin, Michael (1979) *Trapped Within Welfare*. London: Macmillan.

Singer, Isaac Bashevis (1972) *Enemies: A Love Story*. New York: Farrar, Strauss & Giroux.

―――― (1986) *The Penitent*. London: Penguin.

Singer, Peter (1972) 'Famine, affluence and morality', *Philosophy and Public Affairs*, 1, pp. 229–43.

―――― (1976) *Animal Liberation: A New Code of Ethics for Our Treatment of Animals*. London: Jonathan Cape.

―――― (1981) *The Expanding Circle: Ethics and Sociobiology*. Oxford: Clarendon.

―――― (1982) 'Ethics and sociobiology', *Philosophy and Public Affairs*, 11(1), pp. 40–64.

―――― (1984) *Practical Ethics*. Cambridge: Cambridge University Press.

―――― (1985) 'Prologue: ethics and the new animal liberation movement', in Peter Singer (Ed.) *In Defence of Animals*. Oxford: Basil Blackwell, pp. 1–10.

―――― (1993) 'Culture clash sets rite against reason', *The Australian*, 9 June, p. 17.

—— (1995) *How Are We to Live? Ethics in an Age of Self-Interest*. New York: Prometheus.

—— (1997) *Rethinking Life and Death: The Collapse of Our Traditional Ethics*. Melbourne: Text.

Singer, Peter and Mason, Jim (2006) *The Way We Eat: Why Our Food Choices Matter*. Emmaus: Rodale Press.

Siporin, Max (1975) *Introduction to Social Work Practice*. New York: Macmillan.

—— (1982) 'Moral philosophy in social work today', *Social Service Review*, 56, pp. 516–38.

—— (1986) 'Contribution of religious values to social work and the law', *Social Thought*, 12(4), pp. 35–50.

Skinner, B.F. (1973) *Beyond Freedom and Dignity*. Harmondsworth: Penguin.

Smuts, Barbara (1999) 'Reflections', in J.M. Coetzee *The Lives of Animals*. Princeton: Princeton University Press, pp. 107–20.

Solas, John (2002) 'The poverty of postmodern human services', *Australian Social Work*, 55(2), pp. 128–35.

Sorabji, Richard (1993) *Animal Minds and Human Morals: The Origins of the Western Debate*. London: Duckworth.

—— (2006) *Self: Ancient and Modern Insights about Individuality, Life, and Death*. Chicago: University of Chicago Press.

Sorrell, Roger (1988) *St Francis of Assisi and Nature: Tradition and Innovation in Western Christian Attitudes toward the Environment*. Oxford: Oxford University Press.

Soskice, Janet Martin (2004) 'All that is', *Times Literary Supplement*, 9 April, p. 8.

Soule, Michael (1995) 'The social siege of nature', in Michael Soule and Gary Lease (Eds) *Reinventing Nature?: Responses to Postmodern Deconstructionism*. Washington: Island Press, pp. 137–70.

Specht, Harry and Courtney, Mark (1995) *Unfaithful Angels: How Social Work Has Abandoned Its Mission*. New York: Free Press.

Spencer, Nick (2009) *Darwin and God*. London: SPCK.

Spencer, Nick and Alexander, Denis (2009) *Rescuing Darwin*. London: Theos.

Spiegel, Marjorie (1989) *The Dreaded Comparison: Human and Animal Slavery*. New York: Mirror.

Spinoza, Benedictus (n.d.) *Ethics*. London: Heron Books.

Sprigge, Timothy (1979) 'The animal welfare movement and the foundation of ethics', in David Paterson and Richard D. Ryder (Eds) *Animals' Rights: A Symposium*. Fontwell: Centaur, pp. 87–95.

Stalley, R. (1978) 'Non-judgmental attitudes', in Noel Timms and David Watson (Eds) *Philosophy in Social Work*. London: Routledge and Kegan Paul, pp. 91–110.

Steiner, Gary (1998) 'Descartes on the moral status of animals', *Archiv fur Geschichte der Philosophie*, 80(3), pp. 268–91.

—— (2005) *Anthropocentrism and Its Discontents: The Moral Status of Animals in the History of Western Philosophy*. Pittsburgh: University of Pittsburgh Press.

Stevenson, Leslie (1974) *Seven Theories of Human Nature*. Oxford: Clarendon.

—— (Ed.) (1981) *The Study of Human Nature*. Oxford: Oxford University Press.

Stevenson, O. (1971) 'Knowledge for social work', *British Journal of Social Work*, 1(2), pp. 225–37.

Stock, Sarah (2001) 'Aborting one twin should be "an option"', *The Australian*, 29 November, p. 5.

Strean, Herbert (1996) 'Psychoanalytic theory and social work treatment', in Francis J. Turner (Ed.) *Social Work Treatment: Interlocking Theoretical Approaches*. New York: Free Press, pp. 523–54.

Sumner, L.W. (1987) *The Moral Foundation of Rights*. Oxford: Oxford University Press.

Taft, Jessie (1937) 'The relation of function to process in social case work', *Journal of Social Work Process*, 1(1), pp. 1–18.

Tatia, Nathmal (2002) 'The Jain worldview and ecology', in Christopher Chapple (Ed.) *Jainism and Ecology: Nonviolence in the Web of Life*. Cambridge: Harvard University Press, pp. 3–18.

Tawney, R.H. (1930) *The Acquisitive Society*. London: G. Bell & Sons.

—— (1938) *Equality*. London: Allen & Unwin.

—— (1948) *Religion and the Rise of Capitalism*. West Drayton: Penguin.

Taylor, Angus (1999) *Magpies, Monkeys, and Morals: What Philosophers Say about Animal Liberation*. Peterborough: Broadview Press.

Taylor, Charles (1985) *Philosophical Papers Volume 2: Philosophy and the Human Sciences*. Cambridge: Cambridge University Press.

—— (1989) *Sources of the Self: The Making of Modern Identity*. Cambridge: Harvard University Press.

Teichman, Jenny (1985) 'The definition of *person*', *Philosophy*, 60, pp. 175–85.

Tennyson, Alfred Lord (1890) *The Works of Alfred Lord Tennyson*. London: Macmillan.

Terrill, Ross (1974) *R.H. Tawney and His Times: Socialism as Fellowship*. London: Andre Deutsch.

Tester, Keith (1991) *Animals and Society: The Humanity of Animal Rights*. London: Routledge.

Thomas, Keith (1983) *Man and the Natural World: Changing Attitudes in England 1500–1800*. London: Allen Lane.

—— (2009) *The Ends of Life*. Oxford: Oxford University Press.

Thomson, Judith Jarvis (1971) 'A defense of abortion', *Philosophy and Public Affairs*, 1, pp. 47–66.

Thoreau, Henry David (1968) *Walden*. London: Dutton.

—— (1980) *A Week on the Concord and Merrimack Rivers*. Princeton: Princeton University Press.

Tillich, Paul (1962) 'The philosophy of social work', *Social Service Review*, 36, pp. 513–16.

Timms, Noel (1983) *Social Work Values: An Enquiry*. London: Routledge & Kegan Paul.

Titmuss, Richard (1977) 'Who is my stranger?', in Noel Timms and David Watson (Eds) *Talking About Welfare: Readings in Philosophy and Social Policy*. London: Routledge & Kegan Paul, pp. 207–36.

Tobias, Michael (1991) *Life Force: The World of Jainism*. Berkeley: Asian Humanities Press.

Tobler, Helen (Agencies) (2003) 'Foetal eggs "harvest"', *The Australian*, 3 July, p. 3

Todorov, Tzvetan (2000) *Facing the Extreme: Moral Life in the Concentration Camps*. London: Phoenix.

Townsend, Aubrey (1979) 'Radical vegetarians', *Australasian Journal of Philosophy*, 57(1), pp. 85–93.

Toynbee, Arnold and Ikeda, Daisaku (1976) *Choose Life*. London: Oxford University Press.

Trainor, Brian (2002) 'Post-modernism, truth and social work', *Australian Social Work*, 55(3), pp. 204–13.

Turner, E.S. (1992) *All Heaven in a Rage*. Fontwell: Centaur.

Turner, Francis J. (Ed.) (1996a) *Social Work Treatment: Interlocking Theoretical Approaches*. New York: Free Press.

—— (1996b) 'Theory and social work treatment', in Turner, Francis J. (Ed.) *Social Work Treatment: Interlocking Theoretical Approaches*. New York: Free Press, pp. 1–17.

Turner, James (1980) *Reckoning With the Beast: Animals, Pain, and Humanity in the Victorian Mind*. Baltimore: John Hopkins University Press.

Uren, Bill (2002) 'The ethics of stem cell research', *Eureka Street*, 12(10), pp. 9–11.

Vigilante, Joseph (1974) 'Between values and science: education for the profession during a moral crisis *or* is proof truth?', *Journal of Education for Social Work*, 10(3), pp. 107–15.

Walker, Stephen (1983) *Animal Thought*. London: Routledge & Kegan Paul.

Walters, Kerry and Portmess, Lisa (Eds) (1999) *Ethical Vegetarianism from Pythagoras to Peter Singer*. Albany: State University of New York Press.

Warren, Mary (1973) 'On the moral and legal status of abortion', *Monist*, 57(1), pp. 43–61.

—— (1997) *Moral Status: Obligations to Persons and Other Living Things*. Oxford: Clarendon.

Watson, David (1978) 'Social services in a nutshell', in Noel Timms and David Watson (Eds) *Philosophy in Social Work*. London: Routledge & Kegan Paul, pp. 26–49.

—— (1980) *Caring for Strangers: An Introduction to Practical Philosophy for Students of Social Administration*. London: Routledge & Kegan Paul.

Watson, John (1924) *Psychology from the Standpoint of a Behaviorist*. Philadelphia: J.B. Lippincott.

Webb, Mary (1945) *Gone to Earth*. London: Jonathan Cape.

Weber, Thomas (1991) *Conflict Resolution and Gandhian Ethics*. New Delhi: Gandhi Peace Foundation.

Weedon, Chris (1987) *Feminist Practice and Poststructuralist Theory*. Oxford: Basil Blackwell.

Weil, Simone (1952) *Waiting on God*. London: Routledge & Kegan Paul.

—— (1986) *Simone Weil: An Anthology*. (Edited by Sian Miles) New York: Weidenfeld & Nicolson.

—— (2002) *The Need for Roots: Prelude to a Declaration of Duties Towards Mankind*. London: Routledge Classics.

Weinbren, Dan (1994) 'Against all cruelty: the Humanitarian League 1891–1919', *History Workshop Journal*, 38, pp. 86–105.

Weyers, Wolfgang (2007) *The Abuse of Man: An Illustrated History of Dubious Medical Experimentation*. New York: Ardor Scribendi.

White, Lynn (1967) 'The historical roots of our ecologic crisis', *Science*, 155, pp. 1203–7.

Whitman, Walt (1982) *Walt Whitman: Complete Poetry and Collected Prose.* New York: Library of America.

Whittaker, James (1974) *Social Treatment: An Approach to Interpersonal Helping.* Chicago: Aldine.

Wiesel, Elie (1982) *The Town Beyond the Wall.* New York: Schocken.

Wilkes, Ruth (1981) *Social Work with Undervalued Groups.* London: Tavistock.

—— (1985) 'Social work: what kind of profession?', in David Watson (Ed.) *A Code of Ethics for Social Work: The Second Step.* London: Routledge & Kegan Paul, pp. 40–58.

Williams, Bernard (1981) *Philosophical Papers: 1973–1980.* Cambridge: Cambridge University Press.

—— (1985) 'Which slopes are slippery?', in Michael Lockwood (Ed.) *Moral Dilemmas in Medicine.* Oxford: Oxford University Press, pp. 126–37.

Williams, Raymond (1974) 'Social Darwinism', in Jonathan Benthall (Ed.) *The Limits of Human Nature.* New York: Dutton, pp. 115–30.

Williams, Rowan (2000) *Lost Icons: Reflections on Cultural Bereavement.* Edinburgh: T. & T. Clark.

Wilson, E.O. (1975) *Sociobiology: The New Synthesis.* Cambridge: Harvard University Press.

—— (1978) *On Human Nature.* Cambridge: Harvard University Press.

—— (1984) *Biophilia.* Cambridge: Harvard University Press.

Wittgenstein, Ludwig (1953) *Philosophical Investigations.* Oxford: Blackwell.

Wolin, Richard (2004) *The Seduction of Unreason: The Intellectual Romance with Fascism from Nietzche to Postmodernism.* Princeton: Princeton University Press.

Woodroofe, Kathleen (1971) *From Charity to Social Work: In England and the United States.* London: Routledge & Kegan Paul.

Woods, Mary and Hollis, Florence (1990) *Casework: A Psychosocial Process.* New York: Random House.

Wordsworth, William (n.d.) *The Poetical Words of Wordsworth.* London: Frederick Warne.

Wright, Judith (2004) *Birds: Poems.* Canberra: National Library of Australia.

Wynne-Tyson, Jon (Ed.) (1985) *The Extended Circle: A Dictionary of Humane Thought.* Fontwell: Centaur.

Yeo, Stephen (1977) 'A new life: the religion of socialism in Britain, 1883–96', *History Workshop Journal*, 4, pp. 5–56.

Younghusband, Eileen (1964) *Social Work and Social Change.* London: Allen & Unwin.

# Index

subjectivity, 3, 10–12, 20–1, 23–8,
    34–5, 37–8, 42–5, 52, 63, 68, 72,
    75, 78–80, 82, 85, 88, 93, 102–4,
    106, 116, 119, 121, 125–7, 133,
    147–50, 152, 162, 166
  in animals, 3, 11–12, 21, 35, 42–5,
    64, 78–80, 82, 84, 90, 93, 102,
    119, 127, 147–50, 152, 162, 166
  assailing of, 27–9, 34–5, 37–8, 72,
    84, 103–4, 125–6, 131
  independent of language
    possession, 78
  social work, 23, 125–6
  rationality, 11
  *see also* intersubjectivity
subjects-of-a-life, 44–5, 150
Sumner, L.W., 118
Swain, Shurlee, 16

Taft, Jessie, 26
Tatia, Nathmal, 143
Tawney, R.H., 23–4, 32, 119
Taylor, Angus, 43
Taylor, Barbara, 73
Taylor, Charles, 24, 27–8
Teichman, Jenny, 111, 115
Telfer, Elizabeth, 6, 29, 110, 118,
    121–3, 128, 131–5
Tennyson, Alfred Lord, 76
Teresa, St., 123
Terrill, Ross, 25
Tester, Keith, 50–1
Thatcher, Margaret, 27
Thomas, Keith, 7–9, 24, 50,
    75, 80
Thomson, Judith Jarvis, 139–40
Thoreau, Henry David, v, 21, 79
Tillich, Paul, 31
Timms, Noel, 22, 29, 31, 34, 39, 41,
    60, 122, 134
Titmuss, Richard, 47
Tobias, Michael, 143
Tobler, Helen (Agencies), 139
Todorov, Tzvetan, 22, 39, 154
Townsend, Aubrey, 116
Toynbee, Arnold, 7
Trainor, Brian, 38
Turner, E.S., 16
Turner, Francis J., 22–3

Turner, James, 14
Tyack, Peter, 55

uniqueness of human life, 26, 48, 62,
    76, 99, 109, 116, 122, 127, 149,
    161
Uren, Bill, 132
utilitarianism, 15–16, 46, 124, 144,
    148–9, 160–1

value, extrinsic/intrinsic, 42, 107, 111
vegetarianism, 163
Victorian Society for Prevention of
    Cruelty to Animals, 16
  concern for children, 16
Vigilante, Joseph, 27, 34
virtue, 28, 30, 38–9, 55, 61, 137
voluntarism, 28–9, 31, 41

Walker, Stephen, 80
Walters, Kerry, 163
Warren, Mary, 42, 135, 139
Watson, David, 23, 47, 126, 134–8,
    154
Watson, John, 35, 100
Watts, Rob, 119
Webb, Mary, 54
Weber, Thomas, 149
Weedon, Chris, 38
Weil, Simone, 31, 36, 118, 141
Weinbren, Dan, 15
welfare dependency, 119
Welfare state, 69
Western natural philosophy, 8–9
Weyers, Wolfgang, 132
White, Caroline Earle, 17
White, Kenneth, 5
White, Lynn, 7
Whitman, Walt, 30, 98
Whittaker, James, 33
Wiesel, Elie, 48
Wilberforce, William, 14–15
Wilkes, Ruth, 22–4, 26–7, 29, 31, 33,
    35, 104, 112, 122, 124–5, 135,
    151, 165
Williams, Bernard, 112–13
Williams, Raymond, 14, 67
Williams, Rowan, 137, 140
Wilson, E.O., 66–7, 70–2, 77–8, 81,108